Ötzi Man

Ötzi Man

Authors:
Austin Mardon
Marcey Costello
Priyanshu Mahey
Ida Marchese
Terrence Wu
Dasarathy Mutharasan
Ruchira Nandasiri
Ehimen Ogadu
Rosalie Sullivan
David Supina
Armita Yousefi
Yash Joshi

Editor & Project Coordinator:
Stephanie Lazar

Cover Design:
Josh Harnack

Copyright © 2021 by Austin Mardon

All rights reserved. This book or any portion thereof may not be reproduced or used in any manner whatsoever without the express written permission of the publisher except for the use of brief quotations in a book review or scholarly journal.

First Printing: 2021

Typeset and Cover Design by Josh Harnack

ISBN: 978-1-77369-271-5

Golden Meteorite Press
103 11919 82 St NW
Edmonton, AB T5B 2W3
www.goldenmeteoritepress.com

Table of Contents

What is the History Behind the Ötzi Man? 9

Who Discovered Ötzi the Iceman and What Followed the Initial Discovery? 16

What is the Impact of the Discovery of the Ötzi Man? 24

What is the Importance of the Ötzi man? 32

What is the Ötzi Man? 39

What is the Status of the Ötzi Man in the World Today? 46

What Science is Involved in Studying and Analyzing the Ötzi Man? ... 54

How has the Environment Affected the Ötzi Man? What Questions are We Still Asking About the Ötzi Man? 61

What Controversy is There, Surrounding the Ötzi Man? 69

How has the Ötzi Man been Portrayed in Popular Culture? 79

What Future Direction will the Research and Analysis of the Ötzi Man Take? 87

References 95

Chapter 1:
What is the History Behind the Ötzi Man?
by Priyanshu Mahey

Abstract

This chapter touches on the history behind Ötzi and different aspects of his discovery have led to a more detailed recount of not only Ötzi, but also the rest of the bronze age. This chapter begins by introducing Ötzi, discussing his discovery, going over his tools and clothes, then discussing what they reveal about his history. Afterwards, this chapter goes into early predictions on the life of Ötzi, his tattoos, his murder and current thoughts. Ötzi man's history is not only his own personal history but it is the history of all humans, a mirror into the world of the past.

Introduction

History has a lot to offer. The world has existed for over four billion years and humans have only been here for a mere fraction of that time. In order to understand what the rest of that time era before us modern day humans was like, we rely on history in order to understand the past and learn from it. The arduous part of studying history is obtaining significant evidence that is well preserved enough to give us strong insight into the past. The studying of history often requires piecing together poorly aged or scarcely placed artifacts from the past and is often a struggle for historians to figure out. Oftentimes, the biggest issues with studying the past are simply these artifacts. It becomes practically impossible to study the past with little to no evidence and, oftentimes, the artifacts we do obtain are in such poor condition that the only thing they might be good for is display in a museum.

Imagine the joy that came to the face of archeologists and historians when they managed to obtain their hands on Ötzi, an ancient man with his body nearly perfectly intact with clothes and tools. Ötzi was an extraordinarily well preserved ice mummy who gave historians and archaeologists prolific insight into Neolithic life (around 4500 BCE). The fact that it was well preserved gave researchers a pleasantly unbelievable amount of information to work with. Ötzi is the oldest and best preserved mummy discovered and beside him were tools and other ancient artefacts which provided even more insight into ancient life (Maderspacher, 2008).

Story of Ötzi's Discovery

On the nineteenth of september, 1991, Erika and Helmut Simon were scaling the Ötztal Alps. After climbing up Fineilspitze, they attempted to make their way down to the Similaun Hut which was on the lowest part of the mountain ridge connecting Austria and Italy. As they were heading down, they happened to notice a peculiar upper body sticking out at around 3210 m altitude. Curious, they freed the body and reported it to the warden at the hut. At the time, everyone believed it to be the body of a modern man and the longest guess of its age was only 500 years. Initially, everyone assumed it to simply be a man who had died from exhaustion and simply froze to death (Britannica, 2020). Eventually, he was brought to the Institut für Gerichtliche Medizin at the University of Innsbruck where it was found that it was likely he was over 4000 years old. Afterwards, he was placed into the South Tyrol Museum of Archaeology in Bolzano, Italy (Kutschera & Rom, 2000). His discovery was met with genuine shock and happiness by the community and afterwards produced many different studies from experts in many different fields.

Ötzi's Tools and Clothing

Ötzi's corpse was also accompanied with an assortment of memorabilia from the past. In the snow buried next to him, investigators were able to find equipment and clothes. They found clothing items including a fur cap, a belt, shoes, a hide coat, a grass cloak, and leggings all of which were made from tanned leather and grass. Furthermore, they also found spindle whorls, loom weights, a copper axe, flax and vegetable fibers. All of these tools provided a substantial amount of understanding for not only Ötzi's life but the rest of the civilization during the copper age as well (Püntener & Moss, 2010). A lot of what is now known about Ötzi is through the use of radiocarbon dating techniques which enabled researchers to learn everything from the amount of carbon dioxide in the air he took in, to the diet he ate to the fungi that grew on him. A thorough analysis also allowed for a computer to recreate his face and construct a prediction of what he was expected to look like (Kutschera et al., 2000).

Tools provide context and supplementary information, giving insight into the time era. The tools Ötzi had have been very helpful in studying the copper age. Firstly, his one of a kind copper axe, built from leather and copper blade, was no doubt one of the most impressive items they found. This was used as both a tool and a weapon and was undoubtedly a significant discovery. This tool was worn out and clearly used fairly heavily. It is unknown why this axe was taken and why exactly Ötzi had it in the first place (Püntener & Moss, 2010). Besides the axe, there were also arrowheads and a dagger with him. The arrowhead was made of thread and bark. The hypothesis was once that Ötzi had used a recycled blade to make these; however it was found that this was not the case. The true history of this is unclear but it seemed as if Ötzi had recently sharpened the arrowtips without having had a clear reason to use them again. The dagger was another interesting artefact. The dagger was made from a handle of ash wood with a chert blade inserted into it. There was a certain amount of expertise in it's craftsmanship (Wierer et al., 2018).

The clothing was also of interest to many researchers. The clothing used non woven material, relying instead on only tanned leather and grass as well as animal sinews, parts of plants to stitch everything together (Püntener & Moss, 2010). Ötzi's clothing originated from many different places. His hide coat was made from goat and sheep stitched together. His grass cape was made of alpine swamp grass. His leggings, loincloth and belt seemed to be made from strips of domestic goat and sheep hide which were reinforced with additional leather strips. His underwear was made from narrow strips of sheep hide. He had shoes made from string nets from a lime tree bast as well as dry grass which was used for insulation. The outer part of the shoe was made from deer hide. One of the most impressive parts of his clothing is his bearskin cap. Bears are not easy animals to hunt and it is rather impressive that Ötzi was able to have such a cap.

All of these articles of clothing and tools present us with more evidence and potential inputs to develop better theories on Ötzi's history. They gave us an important window into the lives of humans during the bronze age. Looking at Ötzi's tools, it becomes apparent they were heavily used. Their main purpose seemed to have been for planting and cutting. This includes the arrowheads which could have been repurposed sickles. It seemed as if Ötzi did not use daggers for cutting and scraping which is peculiar as it would have been much more efficient. The tools are also beneficial for giving more details on Ötzi's last days as well. Many of his tools were worn out or broken. Some of them seemed as if they were freshly retouched and had been reworked not too long before his death. Looking at the tools as well as the digestive history of the mummy, researchers have been able to piece together the last while before his death. Before his death, the iceman's equipment showed damage and seemed to have been engaged in conflict. This all is linked in to the idea which will be brought later on which suggests that he was murdered (Wierer et al., 2018).

Ötzi's Tattoos

Another interesting thing noted about Ötzi was his tattoos. Tattooing is a fairly ancient practice which is often tied in with culture and religion. In the past, they were often signs of social status and they're tied to hierarchies and provide context on the culture and time period the tattooing took place in. Some of the oldest known tattoos today actually belong to Ötzi who had 61 tattoo marks on his body. He had markings visible all over his body which researchers have continued to study in depth (Deter-Wolf et al., 2016). Beyond his tattoos, radiographic studies show clinical evidence of osteochondrosis and spondylolisthesis which actually have caused medical experts to believe these tattoos were drawn on for medical reasons (Drew, 2017). In the past, it was hypothesized that tattoos were very similar to forms of acupuncture and had therapeutic reasons.

Ötzi's tattoos may have been incredibly common over that time and may indicate many different possible health complications (Zink, 2019). Researchers studied these tattoos and the rest of his body more in depth in order to fully understand what these tattoos meant. They noted that there were several possible musculoskeletal conditions that could have been the prevailing reason behind the tattoos. Some of the prevailing theories are that coronary heart disease or lyme disease were the causes behind his musculoskeletal conditions. These could have led to Ötzi obtaining tattoos in an attempt to be cured of his issues (Kean et al., 2013).

Ötzi's History

Studies have revealed a heavy amount of information about not only Ötzi but the rest of the copper age as well. Using mitochondrial DNA, researchers were able to locate the closest genetic relatedness in order to deduce his background. They found that Ötzi was likely to be variation of European based on how closely related he was to central and Northern European populations (Handt et al., 1994). Researchers later were able to use radiogenic and stable isotopes to better study Ötzi's teeth and bones in order to deduce his habitat range from the days of his youth to his final resting days as an adult. It was found that Ötzi had primarily remained in the valleys of central Europe and his range was restricted around sixty kilometers of his discovery site (Muller, 2003). His tools had indicated that trade and exchange of materials was common even at that time. Additionally, his tools were also very limited, meaning he had to learn how to best make use of the material he had on hand. Many other people at the time would be in very similar conditions to Ötzi and most likely would be just as limited with tools and resources.

Ötzi's Death

Ötzi's death has been a prolific and fruitful field to study. The corpse was exceptionally well preserved and it led researchers to question his death more. One of the things researchers were able to learn about him was the meals he ate and the story associated with the meals by analyzing and extracting the DNA in his intestine. The last food Ötzi ate was primarily red deer, meat, and possibly cereals. Using polymerase chain reaction, researchers found plant remains as well as signs of pollen. Additionally, fungal DNA was also found. Within his stomach, muscle fibers were present which on further inspection seemed to be red deer and domestic goat. From all the evidence, it seems that Ötzi was a hunter/warrior who was traversing through a subalpine coniferous forest of sorts. From there, he feasted on cereals, other plants and meat. Afterwards, closer to his death, he had another meal of red deer meat and potentially cereals (Rollo et al., 2002; Maixner et al., 2018).

An important thing researchers have been trying to figure out is how exactly he died. Scholars have been led towards the conclusion that Ötzi was actually murdered. On his right hand, there is an interesting potential arrow injury which could have led to his death. The wound is particularly deep and it's believed to have been obtained a few days before death. This also suggests he had been attacked mere days before his death (Nerlich et al., 2003). Radiological findings show extra evidence of an arrowhead hitting Ötzi and being the reason for his death. This arrowhead wound was inflicted at his thorax and with computer assistance, it becomes apparent it had suffered a fair bit of damage (Gostner & Vigl, 2002). The arrowhead is then predicted to have damaged some arteries and then it might have led to a deadly hemorrhagic shock which is an extremely likely scenario (Pernter et al., 2007). It is possible that all this fighting is tied together and this was a long and drawn out fight and the wound on Ötzi's hand was likely from defense whereas the arrowhead was an offensive move by his enemy.

All of the evidence points to Ötzi being the earliest known homicide in human history. His death has carved a huge chunk of what is now known about the mysterious copper age. His death will also likely be unsolved leaving it as probably the longest standing unsolved homicide. The only reason researchers have even been able to learn as much as they have now is due to the well preserved nature of the corpse. Had it been preserved any worse, the knowledge would have been lost and researchers would have been left clueless. Even with it's well preserved nature, researchers still have many unsolved questions and as such, much of Ötzi's history remains a mystery. Was he the chief of a tribe warring with a different clan? Had he died chasing his enemies into the mountains? Did he have a family or had he ended up remaining single, remaining by himself? Was he a shepherd who domesticated the lamb and sheep he ate? All of these answers are likely to be buried in the sands of time.

Conclusion

All in all, Ötzi's discovery was a spectacular find. It provided researchers evidence and the material they needed in order to create better predictions about people in the bronze age. Ötzi's death provided a window into not only his life but also the life of other people around at the time. From his tattoos, to his clothes, to the evidence of murder, researchers were able to gain significant information by being able to observe him. The ice had allowed him to remain well-preserved and through scientific analysis, told researchers a lot about the way the iceman had lived. His exceptional condition has served researchers well, providing them with the sample they needed to develop their theories and advance their studies.

Chapter 2:
Who Discovered Ötzi the Iceman and What Followed the Initial Discovery?
by Ida Marchese

Who Discovered Ötzi the Iceman?

The discovery of Ötzi the Iceman is one of a scientific and popular magnitude as the discovery of King Tutankhamen in Egypt (Barfield, 1994). Ötzi, also referred to as the Tyrolean Iceman, is the world's oldest glacier mummy and one of the most extensively researched ancient human remains (Zink & Maixner, 2019).

The body of Ötzi was discovered on September 19, 1991 (Dickson, 2016) on the Schnalstal/Val Senales Valley glacier by two hikers named Helmut and Erika Simon, the prior a retired caretaker (Johnston, 2003). The Simons were from Nuremberg, Germany. The Simons decided to take a shortcut off the mountain trail while descending the Fineilspitze in the Tisenjoch area of the Ötzal Alps. Further along the descent, the pair noticed a brown protrusion in the ice. Approaching the object, they discovered it was a human corpse in which the head, upper torso and arms were visible above the ice (Rosenberg, 2021). The body was entombed in 93m of ice (Collins, 2021). The Simons reported the corpse to a local landlord at a local Similaun mountain refuge after trudging for an hour (Bita, 2005), who then reported the finding to the Similaun police, thinking the body was of a recent death involving a mountain accident (Rosenberg, 2021). The corpse was then taken to Innsbruck for examination (Collins, 2021).

The Simons were fortunate enough to find the Iceman's body as it emerged from the melting glacier before being able to decompose being exposed to the elements (Arie, 2004). Under the circumstances of finding Ötzi shortly after the ice had thawed, he was not exposed to sunlight and wind for an extended period. The fact that the body was so well preserved suggests that Ötzi must have been buried by snowfall shortly after death (Ganesh, 2015).

The Neolithic glacier mummy was found in a flat and rocky terrain of the Tisenjoch. This location is a pass in the Otztal Alps close to the Austrian border in Italy's Schnals/Senales municipality (Fig. 1). The flat local topography and orientation of the rock hollow provide shelter for the body on a pass level. This rocky hollow was crosswise to the ice flow direction, which decreased the influence of the glacial movement. Contrary to steeper areas of the glacier, it is hypothesized that the corpse was not subject to significant ice-flow since much of the soft tissues remained intact with minor deformations (Heiss & Oeggl, 2008). Alongside Ötzi's body were numerous items and tools relatively well preserved. Among these items were a copper axe, flint knives, bow and arrows, sewn birch bark containers and a fragmented net (Barfield, 1994).

It appears that Ötzi reached Tisenjoch on the edge of the mountain decompression, which acted as a 'campground' and then chose the flat slab of stone on which he was found as a resting place. This resting place constitutes Ötzi's discovery location. The 'campground' was indicated by various items found alongside Ötzi's body including various weapons and personal items such as a backpack and clothing (Gleirscher, 2014).
Another theory about Ötzi's discovery location proposed by Heiss and Oeggl is that based on the spread of fragments of bast fibres found around Ötzi suggests that the iceman was moved to the location of his discovery but had not actually died there. Heiss and Oeggl suggest that he was moved along with all of belongings found at the site including his clothing and outer cape, by melting water since the decompression on the mountain slope would have been covered with snow at the time of Ötzi's death (Gleirscher, 2014).

What Followed the Initial Discovery?

Following the initial discovery of Ötzi by Erika and Helmut Simon, both Italian and Austrian authorities were aware of the finding since it was unclear whether the corpse was found on Italian or Austrian soil. Ötzi's body could not be recovered until September 23, 1991, approximately four days after the initial discovery, due to poor weather conditions on the summit.

Since it was an especially warm summer in the Alps in 1991, many other bodies were found along with mountain passes and such Ötzi's discovery at first was not considered abnormal. A warm wind from Africa deposited Saharan sands on the glacial frontier, accelerating their melting. After many decades, several bodies emerged as the snow and ice melted away (Karnitschnig, 2004). Due to this reason, unfortunately, as Ötzi's body was being procured from the ice, many artifacts were unintentionally damaged (Pilø, 2018).

Ötzi's body was excavated using ice picks, ski poles and jackhammers (Ganesh, 2015). During the excavation of Ötzi's body, large sections of flesh from Ötzi's left hip were removed, resulting in damage to the left thigh. Along with other belongings such as a bow broken into pieces and his backpack's ripped frame (Cullen, 2003). Once excavation of the corpse was complete, it was then taken to Innsbruck for examination (Collins, 2021).

Who was Involved in the Recovery and Examination of Ötzi's Remains?

Following Ötzi's discovery in 1991, numerous morphological, molecular and biochemical examinations have been carried out revealing important insights not only about Ötzi's origin and circumstances resulting in his death, but also provides a unique look into what life was like during the Copper Age (Zink & Maixner, 2019).

Among many individuals responsible for the analysis of Ötzi was Dr. Konrad Spindler, who was head of the investigation of Ötzi's remains at the University of Innsbruck. Dr. Spindler, an archaeologist, was the first to determine that the corpse found on the mountain was prehistoric and not a result of a recent death (Pilø, 2018). Spindler is known for creating the 'Disaster Theory' regarding the potential events leading up to Ötzi's death and mummification (Pilø, 2018).

Among examinations of Ötzi's corpse were discoveries of broken bones and apparent previous injuries. The broken ribs played the central role in Spindler's disaster theory, in which it was hypothesized that Ötzi was involved in some physical altercation resulting in the broken ribs. It is hypothesized that the pain of Ötzi's injuries was challenging to bear at the extreme mountainous altitudes, which led to exhaustion, causing him to die where he laid down, presumably to rest (Cullen, 2003).

Spindler examined the body approximately four days after the discovery and quickly recognized the body being "about four thousand years old." Spindler studied Ötzi for about two years before sharing any preliminary findings (Levy, 2008). Spindler wrote his book about the discovery called The Man in the Ice (Levy, 2008) and only after six years was Ötzi returned to South Tyrol to a specially constructed museum in Bolzano (Phillips, 2002).

However, Spindler's approach to studying Ötzi was not well received by some people. According to Brenda Fowler, author of the book "Iceman: Uncovering the Life and Times of a Prehistoric Man Found in an Alpine Glacier", Spindler made researchers sign contracts that would provide him control of information and findings released to benefit the media, who would pay exclusive rights for discoveries. Fowler argues that Spindler elaborated on what he assumed to be the definitive Ötzi story and made claims of other versions being unauthorized (Phillips, 2002). Thirteen additional years of archaeological research have provided great insight into the Iceman (Goodwin, 2004), where the combination of interdisciplinary research has particularly important information about Ötzi's social ranking and death (Gleirscher, 2014).

Interventional Video Tomography (IVT) is one of the many methods used to analyze Ötzi. IVT was used for computer-assisted 3D navigation for ear, nose, throat and cranial surgery. The IVT technology enables researchers to access video images and spatial sensor information. This technology allowed researchers to get samples of the Iceman's mucosa using an endoscopic approach via the maxillary sinus, the nasal cavity and the larynx, which minimized damage to Ötzi's fragile tissues.

Inspection of the sinuses and mucus membrane revealed mucosa typical of a recent cadaver (Thumfart et al., 1997)

Financial Compensation for Discovery

Per Italian law, official recognition as official discoverers of a specific artifact is a prerequisite to be paid compensation (Johnston, 2003). As such, anyone officially recognized as the discoverer of an archeological artifact is entitled up to 25 percent of its value as an incentive to give the artifact to government authorities (Bita, 2005).

According to Italian law, the Simons were entitled to receive up to a quarter of the value of the Iceman's discovery (Arie, 2004). However, Helmut Simon was only recognized as the official discoverer of Ötzi in 2003, nearly 12 years after the discovery. The Ötzi mummy is housed in the South Tyrol Museum of Archaeology. The museum has made approximately €2 million per year for northern Italian authorities since 1998. Previously, Helmut Simon turned down a proposal of €50,000 from Italian authorities many years before his death. Right before Helmut's death, his lawyers were about to launch a case for him to receive a €250,000 reward from Italian authorities for his discovery (Arie, 2004).

Erika and Helmut Simon sued the Italian government for a discovery fee of Ötzi's body (Coakley, 2008). Erika and Helmut endured a fifteen-year legal dispute with the Italian government over a share of profit revenue generated by Ötzi. According to court estimates made by the Simons, local businesses in Bolzano, including restaurants, souvenir shops and hotels, made approximately USD 5.5 million from tourists visiting specifically to see Ötzi. This estimation does not include the amount of income generated from museums, documentaries, souvenirs and online advertising (Pringle, 2009).

The Simons were paid a finders fee of approximately €250,000. Initially, the Simon couple sought a €300,000 payment from the museum, which resulted in an appeal from local officials. The final appeal in Italian court made by the Simons lawyer announced a six-figure settlement; however, the court ultimately ruled a payment to the Simons of €250,000 (Gibbons, 2009).

Until Helmut Simon died in 2004, Helmut continuously battled for financial reward and official recognition from the South Tyrol government. The Simons and the South Tyrol government had been to court twice regarding the case, and in both instances, the Simons lost. The finder's fee awarded to the Simons went to Erika Simon and her two children ("Alpine hikers get finder's fee for Stone Age mummy Ötzi," 2009).

The Simons were named official discoverers in 2003 by a Bolzano court; however, the province petitioned a Court of Appeal after others came forward claiming they had found Ötzi first. Two women came forward claiming they had discovered Ötzi first, followed by claims made by Slovenian actress Magdalena Mohar Jarc. Jarc claimed she came across Ötzi while filming a commercial in the area and accuses the Simons of stealing all accreditations after asking them to take a picture of Ötzi for her (Bita, 2005).

Death of Helmut Simon

On Friday, October 15, 2004, Helmut Simon, without his typical hiking companion Erika, departed from Bad Hofgastein, near Salzburg, up the 2,134m Gamskarkogel peak. The ascent up the mountain should have taken approximately four hours, but half a metre of fresh snow had fallen, and temperatures were close to freezing (Goodwin, 2004). The weather was unfavourable and would have deterred any experienced climber. However, it remains unclear as to why Helmut decided to climb, especially without a tent. Three days following Helmut's departure and disappearance, snowfall of half a metre followed. Over the following days and search and rescue mission, Erika Simon returned to Nuremberg, and the search was called off (Arie, 2004).

Helmut Simon's frozen body was found approximately eight days later under a sheet of snow and ice. Simon fell 300ft to his death while hiking near the location of Ötzi's discovery (McMahon, 2005).

The Ötzi Curse

There exist many curses attached to artifacts and mummies of ancient unfamiliar worlds. Similar to the curses of King Tutankhamun and the

Hope Diamond, it is believed that those involved in disturbing the peace of a resting mummy will be subject to ill fates.

As of the year 2005, seven people associated with the discovery of Ötzi died. Aside from the unexpected death of Helmut Simon, among those who died include Konrad Spindler, Rainer Hoelzl and Tom Loy. Spindler died from complications of multiple sclerosis (Coakley, 2008). Dieter Warnecke, head of the search and rescue team looking for Helmut Simon, died from a heart attack ("'Curse of the Iceman' Linked to Scientist's Death," 2005). Austrian journalist Rainer Hoelzl who filmed the removal of Ötzi's remains from the mountain, died of a brain tumour ("'Curse of the Iceman' Linked to Scientist's Death," 2005). Kurt Fritz, who was one of the first people to see the Iceman, died in an avalanche and was the only member of his company to be struck by the falling rocks (Coakley, 2008). On the way to present a lecture on Ötzi, Rainer Henn, the lead of the forensic investigation on Ötzi, died in a motor vehicle accident ("'Curse of the Iceman' Linked to Scientist's Death", 2005).

Tom Loy completed research regarding Ötzi's DNA in 2005 (Coakley, 2008). Loy was known for revealing four different types of blood on Ötzi's clothing and weapons. This discovery debunked the theory that Ötzi died alone while hunting and was the National Geographic documentary in 2002. Loy died two weeks after completing a book about Ötzi ("'Curse of the Iceman' Linked to Scientist's Death," 2005).

Italian Archaeologist, Professor Bernardino Bagolini, who studied Ötzi at the University of Trento, died of a heart attack in 1995 at the age of 58, the morning after visiting his mother's grave on the first anniversary of her death (Bita, 2005).

Dr. Johan Reinhard, an archaeologist from the United States of America whose notes suggest that Ötzi's burial position on a flat rock between the two highest peaks in the Ötzal mountains at a point illuminated by the summer solstice, suggesting that Ötzi's death involved rituals and could potentially have been an act of sacrifice. Considering Ötzi was dressed elaborately with a bearskin cap, woven accessories and a patchwork leather coat, surrounded by special tools and weapons, implies where Ötzi's body was found was indeed an altar-like structure (Bita, 2005). Some may speculate this could be the reason for Ötzi's curse, as his body was disturbed from his ritualistic sacrificial resting place.

Many rumours among the villages along the Austrian-Italian border propose Helmut Simon may have walked deliberately to his death. However, many other locals fear Ötzi placed a curse on Simon, claiming his life in revenge for disturbing his peace. However, it is essential to note that these claims of the Ötzi curse are unsubstantiated and merely speculation.

Chapter 3:
What is the Impact of the Discovery of the Ötzi Man?

by Armita Yousefi

Due to the growing advancement of technology, it has been assumed that the history of human development has been linear (Bradley, 2009). This school of thought can be attributed to the development of Darwinian evolution and entails that civilization has developed from simple- the prehistoric era- to complex- the modern era. This growing evolutionary model was applied to many secular concepts in the 1800s by French scholars and was clearly expressed with the findings of prehistoric art (Bradley, 2009). For example, when the cave paintings of the Altamira cave were discovered in 1875, many believed that the cave paintings were not prehistoric and were modernly designed due to the complexity of the work (Bradshaw Foundation, 2020). This instance is a significant example of the association of prehistoric people with the lack of sophistication.

However, through the example of complex prehistoric cave paintings, it is clear that the history of human development has not been linear. This assumption can be applied to the revelations that followed the discovery of the Ötzi man. The discovery of this mummified being has been described in depth previously, and in this chapter, the impact of this discovery will be examined. The effect is mainly concerned with the revelations made about the Ötzi man determining the lifestyle of the people living in that era. Researchers have been analyzing the Ötzi man since his discovery and have been providing conclusions about his lifestyle and his community during his life through the technological tools available currently. Initially, there will be a discussion of the Ötzi man's era classification, which will lead to the impact of these revelations on the knowledge currently available about the era of the Ötzi man. These revelations will then be examined in detail based on the tattoo on his body, the equipment carried, his clothing, and the meals eaten before his death. This examination will reveal that the Ötzi man's lifestyle and attributions are anything but simple and that the prehistoric era is complex, similar to society's current status.

Neolithic Era and the emergence of the Bronze Age in modern-day Italy

The classification of prehistoric eras is essential in the scientific study of the Ötzi man and confirms the reservations made about the categories. It is necessary to preface this discussion by noting that the timeline in which different modern-day regions have reached the Neolithic stage in the prehistoric era classifications differs. The prehistoric era can be separated into three stages: the Stone Age, the Bronze Age, and the Iron Age (Kennedy, 2019). The Stone Age consists of three stages: the Paleolithic era, the Mesolithic era, and finally, the Neolithic era (Kennedy, 2019).

During the Stone Age, the behavior of humans, starting from the Paleolithic era to the Neolithic period, has been expressed in many of the artifacts found. One of the most significant changes observed in the Stone Age is the movement from a hunter-gatherer lifestyle to a sedentary lifestyle. During the Paleolithic era, humans would hunt their food and travel long distances to gather their food. The lifestyle was changed during the Neolithic period to cultivate and produce food through agriculture (Kennedy, 2019). With the rise of agriculture, the Bronze Age followed. This transition entailed the beginning of metalwork with the discovery of bronze, copper, and tin alloy (Kennedy, 2019). This era marks the beginnings of advanced tools to produce art, architecture, and clothing as well.

The timeline of the Ötzi man reveals that he most likely lived between 3100 to 3370 B.C.E (Yeung, 2019). This timeline correlates to the late Neolithic era and the emergence of the Bronze Age, commonly known as the Copper Age, in modern-day Italy. The Neolithic period in modern-day Italy began in 6000 B.C.E, and the emergence of the Bronze Age began in 3500 B.C.E (Malone, 2003). This places the Ötzi man in the Bronze Age, and the findings associated with him correlate considerably and further contribute to the findings of anthropologists about prehistoric eras.

The implications of the Ötzi man's tattoos and ingested meal

AA revelation that puzzled the minds of many anthropologists and historians was the markings resembling modern-day tattoos and the ingested meals of the Ötzi man. It has been identified that through incisions made in different parts of the body and the rubbing of charcoal on those incisions, the tattoos observed on the Similaun man were created (Zink et al., 2019). This discovery was perhaps shocking to researchers as the discovery of the tattoos predated the assumed date of the first instance for tattoos. With the initial discovery of these tattoos, the scientific community questioned their purpose. The central question of the purpose was whether these tattoos were decorative or ritualistic or had medicinal value associated with them. The tattoos on the Ötzi man allowed researchers to gain more information about the customs and medical treatments developed during the Bronze Age in Northern Italy, as researchers knew little information before the time of discovery (Zink et al., 2019).

Scientists found 61 tattoos in 19 different regions of the Ötzi man's body (Zink et al., 2019). The tattoos observed on the discovered body were simple: black lines either parallel to each other or forming a cruciform pattern (Zink et al., 2019). Additionally, researchers found the cruciform patterned tattoos on his right knee and left ankle, and the rest were in parallel lines (Zink et al., 2019). Due to the placement of the tattoos in clusters, it was previously noted that this might have suggested early forms of acupuncture (Zink et al., 2019). This hypothesis stemmed from the grouping of the tattoos on the lower back, wrists, ankles, and knees in which those places suffered degeneration and possibly caused pain (Zink et al., 2019).

More recently, a study in 2019 re-evaluated the injuries sustained by the Ötzi man, the placement of the tattoos, and the ingested meals before his death to provide an analysis of the medical customs of his time (Zink et al.). Researchers discovered that the food ingested before his death confirms the Neolithic movement to agricultural means of providing food through the domestication of plants and animals. Furthermore, through the use of advanced technological devices, such as isotope ratios, researchers could indicate that the Ötzi man had been suffering from periods of gastrointestinal problems and degenerative diseases, which in turn caused him pain.

The study hypothesizes that the devices and tools in his tool bag, such as the birch polypore, could have been used as a medical treatment for the pain caused by his diseases. There were two pieces of birch polypore, Piptoporus betulinus, in the leather band of the Ötzi man. This specific fungus has an active substance, agaricine acid, used for anti-inflammatory and antibiotic purposes (Zink et al., 2019). Therefore, this birch polypore could have been helpful for the Ötzi man and his stomach problems medicinally. Furthermore, it could indicate the common use of herbs and plants for their medicinal qualities during this era. Although other applications of this plant, such as tinder, could be plausible, the medicinal properties of this plant are more likely the purpose of this usage (Zink et al., 2019).

Additionally, the ingested pollen and spores from the bracken fern Pteridium aquilinum suggested that the ingestion of this bracken may have been a regular practice for the Ötzi man (Zink et al., 2019). The bracken has been considered a toxic plant due to its carcinogenic properties; however, the carcinogenic part of the plant can be removed with its submergence in water before ingestion (Zink et al., 2019). Interestingly, this bracken has been reported to be used to treat worm infections in the stomach in India and Pakistan in the modern era and has been used in China as well around 3000 years ago (Zink et al., 2019). Therefore, due to the medicinal effects of the bracken, it may be concluded that the ingestion of this bracken by the Ötzi man may have been used to treat his stomach pains (Zink et al., 2019).

There was also a presence of moss species in discovering the foods ingested by the Ötzi man. Two of the moss species consumed have been known to be used as food wrappings, and one has been concluded as unintentionally swallowed with the water that the Ötzi man drank (Zink et al., 2019). The only moss species that he ingested that could have medicinal purposes is Sphagnum, used as wound dressings during the 20th century (Zink et al., 2019). This particular moss species has medicinal properties, such as a high degree of absorbency, that the Ötzi man could have known. When the Iceman suffered from the stab wound on his right hand, he could have possibly used the moss to dress his wound and accidentally ingested the rest of the moss with his last meal (Zink et al., 2019).

Interestingly, the contents found in the stomach of the Ötzi man help support the constant migration patterns around the Bronze Age. For example, traces of the gut bacterium Helicobacter pylori were found in the DNA analysis of the contents in the stomach of the Ötzi man (Greshko, 2016). This bacterium originated from the Asian strain of this bacterium and contained a small fraction of the African strain (Greshko, 2016). This would mean that a large number of migrations from northeast Africa to Europe would have to have taken place before the existence of the Ötzi man due to the transmission rate of this bacterium (Greshko, 2016). Interestingly, the mix of these strains indicates a much larger migration pattern than suggested previously. Thus, a new beginning for research about these migration patterns is much more complex than previously thought (Greshko, 2016).

From the beginning of this discovery, the location of the tattoos indicated that these tattoos were not decorative (Zink et al., 2019). The tattoos are closely linked with the degenerative joint diseases of the Ötzi man, specifically on his knees, ankle, and wrist (Zink et al., 2019). This led to the suggestion that the placement of these tattoos indicates early forms of acupuncture, as 9 out of the 19 groups of tattoos align with the traditional placements of acupuncture points (Zink et al., 2019). A great example of this observed on the Ötzi man's body is the tattoo on his left ankle in the cruciform pattern. The placement of that tattoo is regarded as the "master point" for treating back pain in traditional acupuncture. Researchers have found that this correlates to the changes in the Ötzi man's lumbar and cervical spine that were degenerative (Zink et al., 2019).

Furthermore, other placements of the tattoos that have been more recently found in 2015 on the thoracic region confirm that the prehistoric people could have used these acupuncture points to treat intestinal disorders (Zink et al., 2019). Interestingly, researchers deduced a modern and complex methodology for executing these tattoos, including three steps: local points in the neighborhood of painful areas (locus dolendi therapy), distal points, and masterpoints (Zink et al., 2019). This complex procedure denies the linear movement in historical eras and confirms that prehistoric advancements were complex and similar to modern-day activities.

As previously mentioned, there is no further information about practices in the alpine region during the time of the Ötzi man; thus, the widespread practice of tattoos from that region cannot be confirmed (Zink et al., 2019). However, this does not invalidate the significant impact of this discovery in the scientific community surrounding this era. As observed with the Ötzi man, this practice predates the earliest Chinese acupuncture material and signifies the systemic procedures that contributed to the passage of information through generations (Zink et al., 2019). Although there is much that is unknown about the practice of tattooing during the time of the Ötzi man, such as its efficacy to treat medical conditions, it allows researchers to understand the complexity of organizational structures and advancements in the medical practices that existed during the prehistoric era (Zink et al., 2019). It is highly probable that the Iceman society had complex forms of medical care and treatment as outlined previously. This can be confirmed with the complexity of the tools and clothing discovered with the Ötzi man.

The implications of the Ötzi man's clothing and tools

The clothing and tools discovered with the Ötzi man contribute valuable information about that era to researchers. He was found wearing a loincloth, leggings, a coat, a cloak, a belt, a cap, and shoes (Romey, 2016). The clothing was created from the skins and hides of several different animals. The following chart (figure 1) consists of a summary of the material and the articles of clothing discovered by researchers as outlined on the South Tyrol Museum of Archeology website.

There are fascinating revelations about the clothing on closer examination. The cloak discovered showed signs of long-term usage due to the signs of repair with grass fibers (Romey, 2016). Additionally, the shoes found on the Ötzi man consisted of two layers that would effectively protect him against the harsh conditions in his environment. A modern recreation of these shoes proved the efficacy and complexity of the shoes worn by the Ötzi man (Origjanska, 2017). The current replica of these shoes was tested by a bear-hunter traveling in the Czech Republic and proved to be comfortable and practical in terms of providing protection (Origjanska, 2017). It is essential to note that the hides and skins used were selectively chosen for the clothing pieces due to their practicality (Romey, 2016). These findings prove the complexity of prehistoric culture and reject the linear narrative of history.

The article of clothing	The material used
Loincloth	Sheepskin
Leggings	Goat hide
Coat	Goat and sheep hides sewed together using animal sinew
Cloak	Alpine swine grass woven together
Belt	Calfskin and leather
Hat (with a chinstrap)	Bearskin
Shoes	Inner layer consisting of a bast fiber net stuffed with dry grass and outer layer consisting of deer hide with fur
Quiver	Wild roe deerskin

Figure 1. The articles of clothing found on the Otzi man and the material used to create them

With the scientific and genetic testing done on the articles of clothing, it was found that the sheepskin used tested more genetically similar to modern-day domesticated sheep in Europe than the wilder sheep that would have been predicted to be found during the prehistoric era (Romey, 2016). Additionally, most of his clothing pieces were made from skins of domesticated animals. These findings contribute to the suggested lifestyle change between the Paleolithic era and the Neolithic era from hunter-gatherer to agriculture and the sedentary lifestyle (Romey, 2016). However, there are certain pieces carried by the Ötzi man that were not created from the skin of domesticated animals, such as his quiver and his hat. It is suggested that the Ötzi man occasionally hunted wild animals; however, most of his lifestyle was through agriculture and farming (Romey, 2016). The hunting of the wild animals and the usage of their skin was mainly for practical reasons and used for protection against harsh weather conditions (Chen, 2016).

Conclusion

The spectacular ways the Ötzi man was preserved have allowed researchers in the historical and anthropological field to conduct scientific testing and come to new revelations about the significance of the Iceman. Although the research done on the Ötzi man confirmed the suggested path of human development, it has also created pathways for more research to be done for this period. The movement from the Stone Age to the Bronze Age has been both developed and supported the theories held about the progression from a hunter-gatherer lifestyle to a more sedentary lifestyle. Furthermore, the research done on the tattoos, meals, and the clothing of the Ötzi man has further proved the complexity of human development and has had an impactful change of perspective. Many scientists are pleased with the well-reserved body that has allowed such extensive research to be conducted and the conclusions to be made (Greshko, 2017). Nevertheless, it is crucial to continue to research the implications of the discovery and advance the anthropological and historical knowledge surrounding the significance of the Ötzi man.

Chapter 4:
What is the Importance of the Ötzi man?
by Terrence Wu

Introduction

The Ötzi man is the oldest fully intact mummified corpse of the human race (Hall, 2007). Forensic and genetic investigations have been conducted to investigate the intricacies during his life, such as his habitat, hunting strategies, and possible food sources. This chapter will focus on three major sections for determining the importance and significance of the Alpine Iceman, nicknamed as the Ötzi man, pertaining to evolutionary history, the ice age environmental conditions and genetic analysis.

Evolutionary History

Ötzi is a 5,200 year old Tyrolean Iceman, whose artefacts were found high in the Eastern Ötzal Alps. He is one of the world's oldest glacier mummies, who was found in September 1991. The full corpse has been found in the Italian sector of the Ötztal Alps and preserved in its entirety. These investigations have contributed to increasing the knowledge of Neolithic humankind (Klaus Oeggl et al., 2007; Zink et al., 2019). In 1991, Ötzi was recovered from a glacier in the Austro-Italian Alps where he died in c. 3300 BCE (Emerson, 2019).

His evolution is characterized by the genetic analysis of mitochondrial DNA (mtDNA). The endosymbiont hypothesis explains the origin of mtDNA. The structural and functional diversity of the mitochondria in living cells states that the diverse range of species on our planet today originated from bacterial endosymbionts. Over time, these bacteria integrated themselves into a host cell's nuclear genome. As these cells divide and replicate with newly encoded genetic information, this pathway introduces the development and proliferation of new species. Comparative studies have revealed the similarities and differences between the tandem repeats of both the coding and non-coding portions of the genome (Gray, 1989). Ötzi's mtDNA haplotype is unique to him and has not been detected in contemporary individuals. This genetic feature makes it difficult for scientists to determine his genetic ancestry with high accuracy (Keller et al., 2012). Assembling the pieces obtained from the mtDNA allows scientists to determine the Iceman's phylogenetic position by comparing his genome to large mtDNA sequence databases (Ermini et al., 2008). Small, non-coding RNAs (known as microRNAs) are used as biological markers to measure ancient samples because they are evolutionarily conserved and stable. Analyzing DNA samples provides insights into potential environmental influences that may have had an influence on the genetics of the population (Keller et al., 2017). Evidence derived from paleogenomics is applied to determining the "demographic history of archaic hominids and the relationships between past populations like the Neanderthals and Denisovans and modern humans" (Green et al., 2010; Keller et al., 2017). DNA methylation patterns are used to gain clues and further insight into gene regulation events in ancient humans (Keller et al., 2017). Using this information allows scientists to understand the evolutionary history of Homo sapiens, which has further applications to developing future treatments and therapeutic interventions for a wide range of diseases.

ICE AGE ENVIRONMENTAL CONDITIONS

Due to the high amount of glaciers present in the environment of the Iceman, the retrieved artefacts suggest that Ötzi's corpse was conserved in frozen conditions. Due to global warming and modern environmental conditions, the ice is melting. However, there has been rapid ice formation in the past and further research needs to determine whether some of the original ice covers still exist to this day (Bohleber et al., 2020). Archaeologists examined the Iceman's clothing and equipment to determine the lifestyle of people living in the early Copper Age in the Alpine region. Ancient documentation has been put together to suggest that he was wearing three layers of clothing and sturdy shoes with bearskin soles while trekking through the Alps on his journey (Hall, 2007). Evidence shows that he was able to prepare fire and repair clothes.

His clothing included a cap, leggings, a loincloth, shoes and a coat. They were made from a variety of hunted materials, including different furs and leathers (Zink et al., 2019). He had a good supply of animal resources, including cattle, sheep, goats, brown bears, and roe deers. There are knowledge gaps and some ambiguity in understanding the origin of some of his original leather materials, as the structural features (such as grain pattern, and other biological markers) required for microscopic imaging have decomposed or have been damaged over time. Polymerase chain reaction (PCR)-based genetic analyses have examined the hair shafts that may have been useful in identifying the quality of his clothes (O'Sullivan et al., 2016). He also carried unfinished bows and arrows for hunting. The Iceman also had a large axe made out of an elbow-shaped wooden handle and a finely handcrafted copper blade that was useful for scavenging, hunting and survival (Zink et al., 2019).

Glaciological analysis and micro-radiocarbon analysis indicate that the Alpine summit has been glaciated for about 5,900 years. Radiocarbon dating (Bohleber et al., 2020). Climate change in the west-central European areas may have affected the rapid global climate cooling. It has been speculated that the cooler and wetter environmental conditions may have favoured the rapid burial and preservation of the alpine Iceman. The mid-Holocene climate influenced the palaeoenvironmental conditions that have an impact within human societies and cultural development within central Europe (Magny & Haas, 2004).

Glacier movements are tracked by observing moraines, which are the remaining fragments after the glaciers recede (Kutschera et al., 2017). Glaciers are suitable indicators for climate change as they respond rapidly to changes in temperature and precipitation (Ivy-Ochs et al., 2007). The "Little Ice Age" has an upper glacierization limit, which corresponds to the maximum Holocene glacier elevation. Ever since the Iceman was buried, the ice cover has started to thin out and has now reached the same level as Ötzi's burial site (Baroni & Orombelli, 1996). Following the gradual ice decay, the end moraine is used as a transitional form between the ice-cored moraine and rock glacier. The moraines and glacial bedrocks are dated using in situ-produced cosmogenic nuclides with Accelerator Mass Spectrometry (AMS), also known as radiocarbon ($14C$) dating (Ivy-Ochs et al., 2007). AMS has a higher detection sensitivity that allows radiocarbon dating to be performed using milligrams of carbon, rather than the minimum gram amounts of carbon that are required for beta counting (Bortenschlager & Oeggl, 2012).

At certain glacier points, the cold ice delivers a baseline of at least 5 millennia of Alpine glacier change, which are the markers that are used to identify signs of climate change and predict climate variability over future decades (Bohleber et al., 2020). Topographic measurements of the glacier surfaces using the front moraines from the Little Ice Age have been used to reconstruct the geometry and dimensions of the large lateral moraines. However, this method is not able to determine the thickness variations in the accumulation zone. When analyzing topographic measurements, the assumption is that there will be slight thickness variations that may or may not be accounted for in the calculations, as high elevations decrease linearly with altitude (Vincent et al., 2005).

GENETIC ANALYSIS

A whole-genome study showed that the Iceman had brown eyes and he was also lactose-intolerant (Keller et al., 2012; Zink et al., 2019). Ötzi also has genetic predispositions to several cardiovascular diseases, including coronary heart disease and atherosclerosis. These were identified based on the vascular calcifications that were radiologically sampled in the mummified body (Keller et al., 2012). In addition, there were many pathological conditions that were found within the Iceman. Studies have shown that he may have been plagued with various health conditions, including joint diseases and gastrointestinal problems (Zink et al., 2019). Archaeologists examined enamel and bone samples to determine the contents of his intestine, which can be used to predict his activity over the last few days before his death (Müller et al., 2003). Doctors have inspected his intestines to determine the root of his gastrointestinal problems. They have found that he had a high concentration of eggs from the whipworm parasite leftover, which links to stomach distress (Hall, 2007). Ötzi's intestinal contents were screened for cereal fragments, which reflects the nutritional diet of prehistoric people and was used to determine his last meal (K. Oeggl, 2000). There were remnants of bran-like foods, including wheat, einkorn and barley. There were also small specks of charcoal, which suggests that Ötzi was using these grains to bake primitive bread over an open fire (Hall, 2007). Phylogenetic analyses also show reliable mtDNA sequences from animal DNA, including red deer (Cervus elaphus) and ibex (Capra ibex) (Ermini et al., 2008).

Microbial analyses detected the presence of Borrelia burgdorferi in the Iceman, which is the oldest documentation of this pathogen in humans to date (Keller et al., 2012). B. burgdorferi is the bacteria associated with Lyme disease. Lyme disease affects children and adults, it is characterized by the presence of a red rash shortly following infection. It is very prevalent in rural areas, as they are transmitted by the ixodid tick. It is unknown whether Ötzi showed symptoms of infection during his lifetime. However, the presence of B. burgdorferi is a contributory factor to joint pain and joint damage that was characterized in his corpse. Evidence from x-rays shows osteoarthritic joint abnormalities and spinal damage. It is hypothesized that the Iceman has spinal pain during the later years of his life (Kean et al., 2013).

To overcome some of the barriers, challenges and limitations of characterizing ancient materials, conventional PCR-based Sanger sequencing and 454 pyrosequencing of autosomal DNA techniques have been used. These mixed sequencing genetic procedures have allowed for the complete sequencing of the Iceman's genome. The advantages associated with using high throughput sequencing allows for the targeted enrichment of specific DNA markers that may have disintegrated over time. Identifying these genetic markers allows for scientists to draw links and phylogenies with ancient human and animal populations (Ermini et al., 2008; O'Sullivan et al., 2016).

An in-depth analysis of the skeleton showed that the mineralized skeleton has been well-preserved, as there is significant dehydration of the soft tissues (Murphy et al., 2003). In his skeleton, there are signs of degenerative tissues which indicates that Ötzi may have been plagued with poor health (K. Oeggl, 2000). Several types of trauma were identified in the skeleton, including rib fractures, hairline skull fractures, and compression deformity of the thorax (Murphy et al., 2003). Using bone samples obtained from Ötzi man, DNA extraction and next-generation sequencing was conducted. These genetic analysis techniques were used to reconstruct the nuclear genome of the Iceman. Single-nucleotide polymorphism (SNP) analysis was also conducted to identify specific phenotypes of interests, including lactase persistence (linked to his lactose intolerance), skin pigmentation, and hematological blood groups. Genetic ancestry was also analysed through SNP analysis. There is a heavy presence of tattoos on the mummified body. Research has shown that tattoos were believed to be a form of treatment for lower back pain and degenerative joint disease of the knees, hip, and wrists. Historical studies have shown that tattoos closely resemble an early form of acupuncture (Zink et al., 2019).

It is important to understand the bodily construction of Ötzi to better understand the circumstances of his death (K. Oeggl, 2000). A forensic radiological investigation determined that there was an arrowhead found under the left shoulder of Ötzi. This evidence became a substantial piece of the puzzle to discovering the cause of his death and what happened to him preceding this traumatic event. A histological, high-resolution multi-slice computerized tomography (CT) scan was conducted. A sharpened piece of flint stone was removed from Ötzi's left subclavian artery, near his heart. This suggests that his death could have been due to a major vascular injury, as he must have been suffering from uncontrollable bleeding due to this lethal wound in a major thoracic artery, which led to a rapid and painful death (Gostner & Vigl, 2002; Hall, 2007; Klaus Oeggl et al., 2007). Prehistoric hunters must have been positioned behind him to precisely strike his left shoulder blade, piercing through the bone and one of the major arteries of the heart. This case has become one of the earliest examples of hemorrhagic shock, as Ötzi was rapidly losing blood and there was not enough oxygen reaching his brain. He would have been feeling very faint as he collapsed, lost consciousness and bled out (Hall, 2007).

Isotopic tracing analysis confirms the southern origin of Ötzi, making him a native of the Vinschgau or Schnals valley. This area is located close to where he was found (Müller et al., 2003; Klaus Oeggl et al., 2007). An isotope study was designed to determine the ontogenetics (birthplace, habitat and range) of Ötzi man. The purpose of using radiogenic isotopes allows for the discovery of his local geological environment, due to the decay of long-lived radionuclides. Strontium (Sr) and lead (Pb) are two radioisotopes that were used in conjunction with stable isotopes, oxygen (O) and carbon (C) (Müller et al., 2003). It was found that Ötzi's bones were composed of Sr and Pb isotopes, which matches the composition of Permian volcanoes (Hoogewerff et al., 2001).

Microfossil and macrofossil analyses are used to learn about diet, medicine, and the state of health of a person. This information can be extrapolated to hypothesize the potential causes and reasons for death. Macrofossil analyses are primarily used to determine the types

of plant species that were available in the environment, as well as how they connect and reflect the Neolithic cultivated plants derived from the inner Alpine area. Pollen analysis is also used to determine Ötzi's habitat and the seasonality of the produce consumed, which can be tied to determining the time of death (K. Oeggl, 2000). Archaeobotanists have found plant and pollen fragments that have been used to track Ötzi's last movements. Over 80 distinct species of mosses and liverworts have been identified on and near the Iceman's body. It has been hypothesized that he was using the moss to wrap or preserve food. Coniferous pine pollen was also found in the Iceman's digestive tract. Compiling all of this evidence together suggests that he may have been climbing at higher altitudes where pine trees and ferns can be found in mixed coniferous forests, just before the last few days of his life
(Ermini et al., 2008; Hall, 2007).

Conclusion

The Iceman's discovery presents many interesting case scenarios, as we use this information to learn about the various medical pathologies that Ötzi was plagued with while trekking through the Alps. His corpse is the first fully preserved and intact mummy that we have a clear record to use for further forensic investigations. Although access to study the corpse is limited, therefore further research should look into whether Ötzi was susceptible to any other risk factors that may have had an impact on the quality of his expeditions.

Chapter 5:
What is the Ötzi Man?
by Ruchira Nandasiri

Introduction

A breakthrough in the history was made in the early 90's with the discovery of Ötzi the iceman from the German-speaking Alto Adige or South Tyrol Alps at 3,200 meters above sea level near the Italian-Austrian border (Hess et al., 1998). The isotope analysis of the materials and soil further confirmed that the origin of the Ötzi was from Italy and most of his life was spent on the southern region of Italian-Austrian border and currently the mummy is kept in a special museum in South Tyrol's capital Bolzano (Maderspacher, n.d.). By the time of the discovery, it was reported that the estimated age of the Ötzi the iceman was around 5,100 to 5,300 years based on both the archaeological dating and accelerator mass spectrometric analysis (Rédei, 2008).

Moreover, the carbon analysis further confirmed that Ötzi is the oldest natural mummy found in Europe who is little younger than the other type of mummies discovered from countries including Egypt and Middle East (Maderspacher, n.d.). At the receipt of the dead body of the Ötzi it was found that the whole body was mummified and frozen and reported to be the most ancient human glacier mummified specimen found up to date (Hess et al., 1998). The preservation of the whole body took place with the continuous snow fall following a freeze drying process protecting the corpse from the animals and the scavengers (Dickson et al., 2003).

Further, Ötzi was reported to be older than the Iron Age men from the Danish peat bogs and older even than the Egyptian royal mummies (Dickson et al., 2003). The mummified iceman was given many different nomenclatures by both the historians and archaeologists including the ice mummy, Frozen Fritz, Tyrolean iceman, the 'Man from Hauslabjoch,' the 'Similaun Man', or the Ötzi (Hess et al., 1998; Maderspacher, n.d.).

Discovery of the Ötzi

Up to date our knowledge on the prehistoric times were closely connected with extensive studies in the burial sites. These burial sites provide the up to date information on how the dead were treated whereas very little information is provided from these sites on how they actually lived (Kutschera & Rom, 2000). Moreover, till date no complete corpus was retrieved from the stone age to provide complete understanding with precise proof of life. However, with the discovery of the Ötzi the whole new chapter in history was written providing real life evidence of life stages of the latter part of the stone age or Neolithicum (~4000 – 7000 years ago) (Kutschera & Rom, 2000). As per our knowledge prehistoric men lived about 10,000 years ago lived their lives as hunters and gatherers and limited application of agriculture was involved. However, with the discovery of the Ötzi some solid evidence on the use of agriculture and farming was revealed.

At the time of the discovery Ötzi was lying in an awkward position, with his left arm sticking out to the right, and his right hand trapped under a large stone. Also the archaeologist discovered his gear and some parts of the clothing were scattered around the body (Dickson et al., 2003). Furthermore, the archaeological, genomic and bone analysis confirmed that by the time of the discovery Ötzi was in his mid forties (46 years). Further, the height of the Ötzi was around 159 cm (5 ft 2.5 inches) indicating he was a quite small man. Moreover, at the point of discovery his 12th rib was missing and both 7th and 8th ribs were broken and healed in his lifetime (Dickson et al., 2003). According to Dr. Vanezis from the University of Glasgow, the right rib cage of the Ötzi was deformed and there were possible fractures on both the 3rd and 4th ribs. Dr. Vanezis suggests that these fractures were considered after the death of the Ötzi referring to the early disaster theory (Dickson et al., 2003).

Recurring of thawing and freezing conditions had led to removal of ample amounts of his epidermis, hair, and fingernails. Hence, the in-depth analysis of fingernails illustrated three Beau's lines, which corresponds to the growth of nails. These Beau's lines further demonstrated that Ötzi was severely ill three times in the last six months of his life and, about two months before the death of Ötzi, he was suffering relentlessly and it lasted over a two weeks span (Dickson et al., 2003). Hence, the discovery further stated that the little toe of the left foot of Ötzi indicated the signs of frostbite and most of his teeth were worn off due to his age and the type of diet. Consequently, at the point of discovery only two human fleas were found in his clothes and no lice were seen. This may be due to the shedding of the epidermis of the Ötzi over the period of time (Dickson et al., 2003). Additionally, Dr. Aspöck of the University of Vienna also found that Ötzi had parasite whipworm infestation that leads to enervating diarrhea, thus the severity of the infestation is yet to be discovered (Dickson et al., 2003).

Another important aspect related to the discovery of the Ötzi was related to the autotrophic organisms including diatoms. Usually the drinking water contains different types of diatoms based on its area and location (Kutschera & Rom, 2000). A thorough analysis of the colon contents of Ötzi found there were 24 different types of diatoms ranging from 10 to 50 μm. While the distribution of diatoms in mountain streams changes with altitude and the retention time of the diatoms in the body being few hours the detailed scanning electron microscopic analysis indicated that Ötzi drank water from different altitudes above and below 1500 m before the tragedy (Kutschera & Rom, 2000). Thus, if he drank the water directly from the streams or carried it with him remains unclear to date.

Ötzi - Clothing and Equipment

At the time of retrieval, the researchers found the body of the Ötzi was lying inside a rock hollow preserving the mummified body instead of crunching it while the formation of glaciers (Maderspacher, n.d.). Most interestingly Ötzi man amazed the researchers who made the discovery due to the unique utensils, items, and clothing designs which were associated with him. The discovery of Ötzi led to many speculations and media reports at the time of the retrieval. Konrad Spindler, an archaeologist who was there at first sight hypothesized that Ötzi may have escaped to shelter in the mountains after being wounded in a fight at his home village (Dickson et al., 2003).

The researchers found that his clothing was much more advanced which consisted of a cloak of woven grass over a striped leather jacket, a bearskin cap and goatskin leggings where they showed relatively higher similarity towards the native Americans (Püntener & Moss, 2010). Moreover, spindle whorls and loom weights and some textile fragments of flax and other vegetable fibres were found in his clothing suggesting that weaving of the clothes were taking place in a time before 5000 years back (Püntener & Moss, 2010). Hence, the cap, the coat, and the leggings were created by many cut pieces sewn collectively and the stitching threads were made of animal ligaments/tendons whereas the application of plant materials were much lower. Surprisingly, the footwear of the Ötzi was made out of deer leather and was insulated with grass to withstand the extreme weather conditions and to protect and comfort the feet (Maderspacher, n.d.).

It was also found that many times the clothing material had been repaired skilfully implicating the durability of the material (Püntener & Moss, 2010). Recent analysis of the Ötzi's clothing via mass spectrometry confirmed that samples of his coat and leggings were from the sheep and the upper leather of his moccasins was from cattle (Hollemeyer et al., 2008). These novel findings further confirmed that Ötzi could be a member of a developed society of farmers or ranchers than from a society of hunters (Püntener & Moss, 2010). Also, both the fur and leather items of the Ötzi were intact and tanned, thereby keeping them very stable in harsh weather conditions. During the process of tanning, fixation of tanning agents within the collagen matrix takes place and it was reported that quality of the leather can be improved with the increased degree of fixation (Püntener & Moss, 2010). The use of saponified land animal fats in the process of tanning and as an excellent source of water repellent in the clothing materials many aided the Ötzi during rainy and snowy weather conditions. The advanced knowledge of the Ötzi's on material sciences may be the starting point where the water repellent treatment of leather and fur took place over 5000 year ago. However, the real truth behind this phenomenon is yet to be understood.

Apart from the clothing, Ötzi was carrying several tools and weapons including a flintstone knife and sheath, a copper blade of an axe, and two arrows (Maderspacher, n.d.). The copper tools alongside stone tools found with the Ötzi proved that he belonged to a higher social status. Also, the use of copper in his tools further confirms that at the time of discovery Ötzi was living in the transitional period between stone age and bronze age stated as the copper age (Maderspacher, n.d.). The copper blade of his axe reinforces the above statement. Furthermore, archaeologists found that the stones and metal found near the body of the Ötzi came from both southern and northern regions of the Alps confirming the extensive trade partnerships between the regions during the past (Maderspacher, n.d.). Above, different types of tools have used more than dozens of different plant materials indicating their versatile knowledge on the environment and plant species. Furthermore, these tools and weapons are comparable to the modern-day tools indicating the advanced technologies used during the days of Ötzi.Apart from the clothing, Ötzi was carrying several tools and weapons including a flintstone knife and sheath, a copper blade of an axe, and two arrows (Maderspacher, n.d.). The copper tools alongside stone tools found with the Ötzi proved that he belonged to a higher social status. Also, the use of copper in his tools further confirms that at the time of discovery Ötzi was living in the transitional period between stone age and bronze age stated as the copper age (Maderspacher, n.d.). The copper blade of his axe reinforces the above statement. Furthermore, archaeologists found that the stones and metal found near the body of the Ötzi came from both southern and northern regions of the Alps confirming the extensive trade partnerships between the regions during the past (Maderspacher, n.d.). Above, different types of tools have used more than dozens of different plant materials indicating their versatile knowledge on the environment and plant species. Furthermore, these tools and weapons are comparable to the modern-day tools indicating the advanced technologies used during the days of Ötzi.

Ötzi – Cultural Impact

The researchers found that the content of the stomach of the mummified body of the Ötzi contained primitive types of cultivated wheat which further illustrated the agricultural and farming knowledge of the Ötzi's (Maderspacher, n.d.). The researchers have further reported that the bran of the primitive wheat called einkorn was also found in the content of his stomach. These wheat particles seem to be fine and well grounded making into a flour implementing the advanced technological aspects of the food processing (Dickson et al., 2003). In addition, wheat contents of the stomach further contained game meat and dried sloes indicating the Ötzi's were an omnivorous group who consumed both plant and animal materials (Maderspacher, n.d.). The microscopic and DNA analysis of the debris of stomach content showed that he was eating both plant materials and meat of both red deer and alpine ibex (wild goat). The discovery of ibex neck bone fragments further confirms the omnivorous feeding habits of the Ötzi (Dickson et al., 2003). Carbon 13 and Nitrogen 15 isotopic analysis also verified that around 30% of his dietary nitrogen was attained by animal sources and the remaining 70% by the plant sources. Moreover, the researchers also found that he was carrying a type of mushroom which had antibacterial properties indicating the knowledge on medicine by the Ötzi's.

Further, the body of the Ötzi was covered with several tattoos in his body including wrist and in other places of his body. These tattoos were made by making fine incisions in the skin followed by rubbing charcoal into the incisions (Kristensen, 2019). Further, these cuts were primarily found on the body where it is more vulnerable to pains and injuries and it was further found that these tattoos were used to treat many diseases including joint pain and arthritis (Kristensen, 2019). These incisions were further visible under the missing epidermis indicating the technology applied towards making the tattoos. Further, it was reported that mostly these tattoos were used as a therapeutic purpose rather than recreational. The presence of incisions near to Chinese acupuncture points and places closer to joints of arthritis patients including lower spine, right knee and ankle confirms the therapeutic properties of the tattoos (Dickson et al., 2003). However, the X-ray reports analyzed by researchers Vanezis and Tagliaro of the University of Rome confirmed that Ötzi was not suffering from arthritis.

Ötzi - Neolithic Life

The researchers have conducted a detailed genomic analysis to identify the relationship of the Ötzi man to common day humans. The analysis found that the one DNA sequence of the hypervariable region of the mitochondrial DNA was similar to the people of central and north European populations (Rédei, 2008). Further, the complete mitochondrial genome analysis confirmed that Ötzi belonged to an ethnic group called haplogroup K associated with the Western European region. However, two mutations found within his genome made him special and unique compared to the current European community (Maderspacher, n.d.). Further genomic analysis revealed that Ötzi had four different types of blood on his clothing indicating the cause of his death was connected to a deadly fight with other people. The cause of the death was reported to be due to an arrow flint by further X-ray analysis (Rédei, 2008). The X-ray analysis proved that an arrowhead was found under his left shoulder blade where the offenders may have pulled the shaft from the body. The archaeologists additionally uncovered that arrowhead pierced his subclavian artery leading towards the loss of blood which caused his death. Hence, the X-ray analysis reinforced that Ötzi also had a severe blow to his head prior to his death (Maderspacher, n.d.).

Conclusion

Today, three decades after the discovery of the iceman, the oldest and best-preserved human body still contains a lot of information hidden from its readers. Up to date how the Ötzi ended up in the high Alps and details about his generation is an unsolved mystery. Furthermore, the genomic and detailed carbon isotopic analysis of plant remains of the pollen, seeds, mosses, and fungi found both inside and outside the body of the Ötzi have revealed many shocking details. Hence, a detailed analysis and autopsy would be really challenging as the body itself is mummified and may lead towards further destruction. Therefore, we may never know the real reason for the death of Ötzi, or what exactly he was doing in the high Alps and how exactly he died. With these limitations in hand, it's sometimes hard to exclude some theories including that Ötzi died elsewhere and was carried to the hollow for the burial till the hikers found him 5000 years later. However, we could conclude that the discovery of Ötzi has made a huge impact on the history of humankind. Keeping all his secrets to himself, Ötzi today lies in a specially built chamber at the South Tyrol Museum of Archaeology in Bolzano, Italy, where they keep him at a temperature of -6oC with 99% humidity further preserving his past for future generations.

Chapter 6:

What is the Status of the Ötzi Man in the World Today?

by Rosalie Sullivan

Introduction:

What is the status of the Ötziman in the world today?

In the year 1991, scientists and archeologists were baffled with the discovery of the Ötzi Man. The Ötzi Man was a prehistoric Iceman that "had been naturally mummified by the ice [he was found in,] and kept in amazing conditions for approximately 5, 300 years" (Rosenberg, 2020). He was an astounding historical and archaeological find, and his frozen body offered much insight into "the life of Copper Age Europeans" (Rosenberg, 2020). The Ötzi Man was "nicknamed after the mountain range where he was found" (Kutschera et al., 2000), and the prehistoric Iceman would become one of the greatest archaeological finds of all time.

In the beginning, however, no one "suspected that the [frozen] body could be from prehistoric times" (Kutschera et al., 2000). When the body was found by two hikers exploring the icy mountain ridge, they did not realize that they had uncovered one of the greatest scientist discoveries of all time. After all, accidents were quite common in the Ötztal Alps, and people have been known to occasionally go missing in the snowy mountains (Kutschera et al., 2000). When the body was being uncovered and removed from the ice, the people working with the Iceman assumed "that the body might be as old as 500 years" (Kutschera et al., 2000). This educated guess, however, was incorrect. "Four days after the discovery, the body was freed from the ice" (Kutschera et al., 2000), and scientists and archaeologists began to realize how truly old the Iceman was. The Ötzi Man was a prehistoric Iceman, an organic relic of the past, and they were astounded. Nothing quite like the Ötzi Man had ever been discovered before, and the world was completely shocked. The Ötzi Man quickly became a well-known scientific and archaeological discovery, and studies were done in order to study the prehistoric Iceman further. The Iceman was able to offer unique and intriguing information about the Copper Age of Europeans (Rosenberg, 2020), and the entire world was sitting on the edge of their seats. The discovery of the Ötzi Man was truly like nothing else.

In recent years, however, more and more studies have been conducted regarding the Ötzi Man. The Ötzi Man is still a hot topic of interest for many scientists and archaeologists, and the Iceman "has been the subject of intensive research ever since [his discovery]" (South Tyrol Museum of Archaeology, 2021). In the world today, the status of the Ötzi Man is constantly shifting and changing, as more and more information about the prehistoric Iceman is unraveled and discovered. Going forward in the future, the conversation and discussions surrounding the Ötzi Man will only continue to grow, as our information about the Iceman grows as well.

In our modern world, there have been quite a few scandals and legal disputes surrounding the Ötzi Man as well; along with lots of drama. The great scientific and archaeological find most certainly came with its own fair share of drama and tension. The drama and legal disputes surrounding the Ötzi Man stirred up quite a few rumors and problems, but they were dealt with accordingly in the end (Walker, 2013). The history of the Ötzi Man is a fascinating one, and the status of the Ötzi Man in the world today is just as interesting. The Ötzi Man is truly one of the greatest scientific and archaeological discoveries of all time.

Where is the Ötzi Man Now?

Since the year 1998, the Ötzi Man has "been on display at the South Tyrol Museum of Archaeology in Bolzano, Italy" (Yeung, 2019). For the past thirty years, the Iceman's location has not changed from that spot, and he remains present at the Italian museum today. Even though other museums have requested access to the prehistoric Iceman, wanting to display him as well, the South Tyrol Museum refuses to part with him

(Yeung, 2019). The idea of letting the Ötzi Man get taken away is an unsettling one for the museum. The Italian museum, after all, makes a lot of money off of the prehistoric Iceman and his exhibit.

The South Tyrol Museum's "public display of Ötzi's remains in the museum brings in a significant number of tourists...each year" (Walker, 2013). The exhibit also brings in tourists not only to the museum, but to the city as well (Walker, 2013). Thanks to the prehistoric Iceman, the entire city is able to experience a boom in business, profit, and tourism. The museum alone makes an estimated 3.2 million US dollars each year in admission, and other businesses in the city are expected to experience higher profits as well (Walker, 2013). Most local businesses in the area will sell an "endless variety of merchandise [of the Ötzi Man]" (Walker, 2013). This merchandise can range from T-shirts to coffee mugs, and even Brad Pitt has a tattoo of the Iceman on his left arm (Walker, 2013). The Ötzi Man's incredible discovery is advertised everywhere in Bolzano, Italy, and his impact on the city is hard to ignore. Undoubtedly, the prehistoric Iceman and his display at the museum brings "an enviable opportunity [for the entire city] to profit economically" (Walker, 2013).

The Italian museum also possesses the Ötzi Man's clothing and equipment as well, which are also on display in the museum's exhibit (South Tyrol Museum of Archaeology, 2021). The clothing and equipment has been studied for many years, proving to be just as interesting and informative as the Iceman himself. Every part of the Ötzi Man is invaluable, and the prehistoric Iceman will continue to fascinate scientists and archaeologists for generations to come. The information about the Copper Age of Europeans that is being discovered through the Iceman is beneficial, and it allows us to develop a better understanding and picture of the time period (Rosenberg, 2020). The Ötzi man, along with his clothing and equipment, is truly a great scientific study that will continue to further scientific and archaeological research on the Copper Age of Europeans. The Iceman will also offer insight into our own biology—how we, ourselves, have changed throughout the ages.

The prehistoric Iceman before getting transferred to the South Tyrol Museum, however, was present at the Austrian University of Innsbruck (Walker, 2013). Scientists and archaeologists studied the Iceman's corpse for seven years at the university, and most of the scientific breakthroughs were made at the Austrian school (Kutschera et al., 2000). After seven years of research, however, "numerous investigations [about the Iceman were] published in a series of monographs" (Kutschera et al., 2000). These publications astounded the public, and people were incredibly intrigued about the prehistoric Iceman—wondering what he could teach them about the past.

Going forward in the future, the Ötzi Man is expected to stay at the South Tyrol Museum for all foreseeable time. The prehistoric Iceman will not be moved from his home in the museum, from his exhibit, and research about him will continue to be conducted.

How is the Ötzi Man being preserved today?

In the present today, the Ötzi Man is being preserved at the South Tyrol Museum "in a specially devised cold cell [that] can be viewed through a small window" (South Tyrol Museum of Archaeology, 2021). The ice that the Ötzi Man had been encased in in the Ötztal Alps "led to [the] exceedingly good preservation [of his body and organic material]" (Oeggl, 2009). The Ötztal Alps' snow, frost, and ice, protected the body from "decay and damage from carrion feeders." (Oeggl, 2009), and because of this, the Ötzi Man's corpse was able to stay intact for many years. The prehistoric Iceman's "mummification had only been possible because the body was exposed to frost and [a cold climate]" (Oeggl, 2009). The Ötztal Alps' frigid temperature and climate allowed for the prehistoric Iceman's body to survive for thousands of years, and the South Tyrol Museum now tries to replicate these conditions.

The South Tyrol Museum, today in the present, takes care of Ötzi the prehistoric Iceman, and provides him with intense and frequent care. The Italian museum ensures that the Ötzi Man has regular multiple check ups each month (Hallett, 2018), so he is kept in pristine condition. Oliver Peschel, a scientist who works for the museum, takes care of the prehistoric Iceman and preserves him with a special misting treatment (Hallett, 2018). It is the scientist's job to ensure that the Ötzi Man is kept in a place with no light—a place where the Iceman can be kept moist, cold, and comfortable (Hallett, 2018). The misting treatment takes approximately fifteen to thirty minutes to complete, and the treatment is done once every eight weeks (Hallett, 2018). Peschel, when performing the misting treatment, keeps an "eye out for any discoloration and evidence of bacterial or fungal growth" (Hallett, 2018). The prehistoric Iceman, after all, needs to be kept in the best condition possible, and any threat to his well being needs to be eliminated.

When the prehistoric Iceman is being put on display, however, and not being examined privately, he is "lying faceup on a large glass plate, his left arm bent across his body" (Hallett, 2018). The display he is kept in is made out of large "blocks of ice, kind of like an igloo" (Hallett, 2018). The ice walls encase Ötzi the Iceman and keep his body nice and cold—well preserved. "Ötzi himself is [also] covered in a thin layer of ice" (Hallett, 2018), so every part of his body can truly be protected from the harsh outside conditions. The Iceman is "safely stored in a glass vitrine with controlled temperature (−6°C) and humidity (98%) at glacier-like conditions" (Kutschera et al., 2000). Every part of the Iceman's display case is regulated so Ötzi can have the best preservation possible. The Italian museum takes care of the prehistoric Iceman to the best of their abilities, and Ötzi is kept in nice cold conditions.

Even though the scientists work hard at preserving the Ötzi Man, however, the prehistoric Iceman still "loses about 2 grams of water weight each day" (Hallett, 2018). Preserving the Iceman is a difficult task, especially when the Ötzi Man is determined to decay—to rot away. Oliver Peschel adds more water and ice to the Ötzi Man on a regular basis, but the Iceman always manages to shed the ice eventually (Hallett, 2018). Taking care of the prehistoric Iceman is no doubt a difficult task, but it is one that the South Tyrol Museum loves doing.

The South Tyrol Museum, however, is planning on renovating and expanding the museum and the Iceman's exhibit (Hallett, 2018). However, since the museum will be changing, the conditions preserving Ötzi the Iceman will need to be changed as well. A few different options have been discussed, but one has not been decided yet. The first option is that the prehistoric Iceman would be encased and suspended in a giant block of ice (Hallett, 2018). The ice block would preserve him and leave him visible to tourists and visitors, but it would make scientific examinations much more difficult (Hallett, 2018). The second idea was that the museum would switch "from an oxygen atmosphere to a nitrogen one, which would be less friendly to bacteria" (Hallett, 2018). This would help with preserving the prehistoric Iceman. The third and last option was for them to vacuum-seal Ötzi, but the idea was ultimately disregarded in favour of the other two options (Hallett, 2018). When the museum is renovated and changed, one of these options will be decided upon and used in the preservation of the prehistoric Iceman.

In the end, Ötzi the Iceman is kept in cold conditions at the South Tyrol Museum, and he undergoes a variety of treatments that maintain his condition. He is cared for greatly, and the scientists absolutely adore him. The preservation of Ötzi is intricate and intense, and the lengths the museum will go for him knows no bounds.

What was the Ötzi Man legal dispute?

The Ötzi Man legal dispute was a problem that came about with the discovery of the prehistoric Iceman. Erika and Helmut Simon, the two hikers who discovered the Iceman, "demanded recognition of their role in discovering Ötzi in the form of a 'finder's fee'" (Walker, 2013). This 'finder's fee' that they were arguing for meant that they would receive twenty five percent of all money that Ötzi made (Walker, 2013). The two hikers were "eager to claim their share of [the Iceman's] fame" (Walker, 2013), and they both wanted to make some money off of Ötzi. In 1994, three years after Ötzi's discovery, Italian authorities offered the hikers a reward of ten million lire for the role they played, but the Simons rejected this offer (Wikipedia, 2021). Later on, in 2003, the Simons filed a lawsuit where they asked to be recognized as the official discoverers of the prehistoric Iceman (Wikipedia, 2021). The court, in November 2003, approved of their claim and sided with the Simons. Near the end of 2003, however, the Simons asked for US$300,000 as their fee, and the provincial government decided to appeal against them (Wikipedia, 2021).

After all of these incidents occurred, however, two more people came forward and claimed that they were "part of the same mountaineering party that came across Ötzi and discovered the body first" (Wikipedia, 2021). Magdalena Mohar Jarc claimed "that she discovered the corpse first after falling into a crevice, and shortly after returning to a mountain hut, asked Helmut Simon to take photographs of Ötzi" (Wikipedia, 2021). The second woman was Sandra Nemet who argued that "she found the corpse before Helmut and Erika Simon, and that she spat on Ötzi to make sure that her DNA would be found on the body later" (Wikipedia, 2021). The woman even requested for a DNA test on the remains, so she could prove the validity in her statement, but experts denied her (Wikipedia, 2021). These two different legal cases "angered Mrs. Simon, who alleged that neither woman was present on the mountain that day" (Wikipedia, 2021). The two legal claims, however, never went anywhere, and the two women were never validated by the court or the government.

In 2008, however, the Simons were finally given a monetary reward for the discovery of Ötzi the Iceman—they were rewarded with €150,000 (Wikipedia, 2021). The legal dispute had finally been settled, and the world could move on from this drama.

There was, however, another legal dispute that continued to rage on longer than the Simons' dispute. This was the dispute between Austria and Italy; the two countries were arguing over "who would [get to] own and permanently house the remains [of Ötzi]" (Walker, 2013). When Ötzi was first discovered, he was taken to the Austrian University of Innsbruck—the Iceman was then studied at the University for quite a few years (Walker, 2013). Later on, however, "it was established that the discovery location [of Ötzi] was actually about one hundred meters on the Italian side of the border" (Walker, 2013), and the Italian government began arguing for the Iceman to be transferred into their custody. In the end, Italy's claim to the Iceman prevailed and they gained ownership of Ötzi.

The legal disputes involving Ötzi the Iceman were complicated and long processes that took many years. The arguing went back and forth for quite awhile, but they have all been settled now in the present day.

Conclusion

In conclusion, Ötzi the Iceman was discovered in 1991 and ever since his discovery, the world has been baffled by the prehistoric Iceman. Ötzi has astounded scientists and archaeologists, and in recent years—more and more studies have been conducted regarding him (South Tyrol Museum of Archaeology, 2021). Going forward in the future, it is expected that the dialogue surrounding Ötzi the Iceman will continue to shift and change as we learn more about him.

Currently, the prehistoric Iceman is being kept in the South Tyrol Museum in Italy, and the surrounding area is able to make a lot of money off of his presence there (Walker, 2013). The Iceman, however, did use to be present at the Austrian University of Innsbruck before he was moved to the museum for further research (Walker, 2013).

Going forward in the future, it is expected that Ötzi will continue to remain at the museum while being contained in a cold display case that is able to regulate his temperature and preserve his body (South Tyrol Museum of Archaeology, 2021). There are, however, problems with preserving the Iceman, and his preservation is not simple and easy. The scientists do work hard at preserving Ötzi, though, and they work long hours with the Iceman. The museum tries its absolute best to study and care for the prehistoric Iceman. Recently, however, the museum has decided to renovate itself and it's exhibits, and the preservation method used on Ötzi the Iceman is expected to change (Hallett, 2018). They have developed three potential methods they might use on the Iceman, but it has not been decided which method will actually be enforced (Hallett, 2018). No matter what they decide upon, however, the prehistoric Iceman will continue to be preserved and well looked after.

The Ötzi Man legal dispute was a problem that came about with the discovery of the prehistoric Iceman. Erika and Helmut Simons, the two hikers who found Ötzi, demanded compensation and a 'finder's fee' for the role they played in the Iceman's discovery (Walker, 2013). After those two, two women came forward claiming that they also were present when Ötzi was discovered, but these claims have never been validated (Wikipedia, 2021). Eventually the Simons were offered monetary compensation for their role in Ötzi's discovery, and the legal dispute was put to rest (Wikipedia, 2021). There was, however, another legal dispute that arose earlier on—the legal dispute between Austria and Italy. Austria, in the beginning, had access to Ötzi the Iceman and was studying him at one of their universities, but when it was realized that Ötzi had actually been on the Italian side of the border—they lost custody of the Iceman to Italy (Walker, 2013). Italy, today in the present, now harbours and cares for the Iceman (Walker, 2013).

Chapter 7:
What Science is Involved in Studying and Analyzing the Ötzi Man?
by Marcey Costello

Copious amounts of research have been conducted by various organizations to understand everything they can about the Ötzi Man, also known as the Iceman, Ötzi the Iceman, or simply Ötzi. Such "extensive scientific studies" include "radiology and computed tomography [CT], histology, isotope analysis, paleobotany, and genetic analysis" (Zink et al., 2014, p. 204). "Histology" is the study of the structure of microscopic tissues and organs (Oxford University Press, 2021). Isotopes are elements that have the same atomic number and chemical properties as a specific element but have different mass numbers because they have a different amount of neutrons in their nuclei (Oxford University Press, 2021). Thus, "isotope analysis" studies the various isotopes within something, in this case, within Ötzi. "Based on stable isotope analysis, it was shown that the Iceman grew up and lived the last years before his death in different valleys in the southern region of the Alps" (Zink et al., 2014, p. 204). Moreover, "paleobotany" is "[t]he branch of botany that deals with extinct and fossil plants" (Oxford University Press, 2021). "A paleobotanical study and pollen analyses of samples removed from [Ötzi's] intestines have provided important insights into his nutrition, his last itinerary, and the season of his death in late spring" (Zink et al., 2014, pp. 204-205).

More recently, the following have been used to study Ötzi: "nanotechnological analysis of soft tissue and bone samples, spectroscopy of blood remnants in his clothing, a reevaluation of radiological data, and a detailed genetic analysis of the nuclear genome" (Zink et al., 2014, p. 204). The "[c]omputed tomography...scans demonstrate the presence of healed rib fractures, degenerative arthritis, some degree of vascular calcification, and the presence of an arrowhead in his left shoulder" (Zink et al., 2014, p. 205). The CT also showed "that the Iceman suffered from a severe brain injury that could have been caused shortly after or before the deadly arrowshot" (Zink et al., 2014, p. 205). "Degenerative arthritis is a condition that involves the chronic breakdown of cartilage in the joints leading to painful joint inflammation" (Spine-health, n.d.). Additionally, "vascular calcification" is "mineral deposits on the walls of your arteries and veins" (University of Pittsburgh Medical Center, 2018). A more recent CT performed in 2007 "clearly demonstrated that the arrowhead had lacerated the left subclavian artery, likely leading to a rapid, deadly hemorrhagic shock" (Zink et al., 2014, p. 205). The "left subclavian artery" runs just under the left collarbone and provides blood to the left arm (Aghoghovwia, B., 2020). Cannon explains that "[h]emorrhagic shock is [when]...severe blood loss leads to inadequate oxygen delivery at the cellular level. If hemorrhage continues unchecked, death quickly follows" (2018, p. 370). Furthermore, "[i]n a recent reevaluation of the CT scans, further details on the life and death of the Iceman were revealed, including the presence of gallbladder stones and a completely full stomach" (Zink et al., 2014, p. 205).

As previously mentioned, Ötzi showed signs of "some degree of vascular calcification" which is "a clinical marker for atherosclerosis" (Zink et al., 2014, p. 205; Abedin, et al., 2004, p. 1167). In other words, vascular calcification is a sign of atherosclerosis. "Atherosclerosis is a disease in which plaque builds up inside...arteries" (Atherosclerosis, n.d.). "Plaque is made up of fat, cholesterol, calcium, and other substances found in the blood. Over time, plaque hardens and narrows your arteries. This limits the flow of oxygen-rich blood to your organs and other parts of your body" (Atherosclerosis, n.d.). Thomas et al. explain, "Using x-ray computed tomography (CT), the investigators contributing to this issue of Global Heart have yet to find a culture that did not have pre-clinical atherosclerosis" (2014, p. 185). "The remarkable preservation of Ötzi... allowed [researchers] to evaluate...phenotype...[and] genotype of this naturally preserved mummy" (Thomas et al., 2014, p. 185). Essentially, the researchers were able to study Ötzi's physical characteristics and his genetic material. "Despite Ötzi's active lifestyle as an alpine hunter-gatherer, atherosclerosis developed in his carotids, aorta, and iliac artery" (Thomas et al., 2014, p. 185). The researchers "identified that Ötzi...harbored many of the alleles that predispose to atherosclerosis during present day" (Thomas et al., 2014, p. 185). "Ötzi, like most other mummies, died before his atherosclerosis could have become clinically manifest [i.e. before the symptoms of the disease affected his life]" (Thomas et al., 2014, p. 185).

Zink et al. also used the "genome approach" to learn more about Ötzi including "that he likely had brown eyes, in contrast to the previous assumption that he had blue eyes" (2014, p. 205). "It could be further shown that [he] was lactose intolerant, a characteristic that is among the most significant genetic traits connected with the beginning of agriculture in Europe" (Zink et al., 2014, p. 205). "Genetically, he most closely [resembled] early European farmers" (Maixner et al., 2016, p. 163).

Additionally, they discovered "indications for recent common ancestry between the Iceman and present-day inhabitants of the area near the Tyrrhenian Sea" (Zink et al., 2014, p. 205). Thus, not only did the "genome approach" allow researchers to discover Ötzi's atherosclerosis, but they were also able to learn even more details about him that they could connect with more recent events and people (Zink et al., 2014, p. 204). "The most intriguing finding was that the Iceman showed a strong genetic predisposition for increased risk for coronary heart disease (CHD)" (Zink et al., 2014, p. 205). "This is of particular interest as the CT scans of the Iceman already had revealed major calcification in carotid arteries, distal aorta, and right iliac artery, which are strong signs of generalized atherosclerotic disease" (Zink et al., 2014, p. 205). In other words, "[t]he genetic predisposition could have significantly contributed to the development of the arterial calcifications" (Zink et al., 2014, pp. 205-206). Zink et al. state, "[T]he Iceman is the only ancient human remain in which a genetic predisposition for cardiovascular disease has been detected" (2014, p. 206).

"Microbiologist Frank Maixner...and an international team of scientists have recently discovered and sequenced the DNA of an extinct strain of gut bacteria,...from...inside [Ötzi]" (Milliken, 2016). The bacteria "Helicobacter pylori" "is no run-of-the-mill gut microbe, and the strain found in Ötzi offers tantalizing clues about the geography and migratory paths taken by early humans" (Milliken, 2016). Helicobacter pylori, or H. pylori, is "[n]estled on the acidic walls of half of the human stomachs on

Earth" and "has evolved with Homo sapiens for at least 100,000 years to become specifically attuned to the acidic environments of the human stomach" (Milliken, 2016). "It was present in the first modern humans to walk out of Africa and has since followed a near identical geographic distribution around the globe" and "[a]s a result, H. pylori has become a marker of the history of human migration and dispersal around the world" (Milliken, 2016). Milliken explains that "H. pylori is not particularly beneficial (although it could be important in overall gut ecology), but it rarely is harmful...[a]t least in the modern age" (2016). "Of the 50 percent of humans that carry it, only 10 percent develop a disease that causes ulcers or gastric cancer" but rather "[u]nfortunately for the Iceman, the now extinct strain of H. pylori that he carried was virulent, and showed signs of having caused an inflammatory response from his body" (Milliken, 2016). In order to collect a sample from Ötzi's "desiccated stomach", the scientists used a previous incision spot to avoid making a new one (Milliken, 2016). "To find the H. pylori, the researchers...[collected] DNA samples from everything that was in the Iceman's stomach. Unfortunately that generated several hundred gigabytes of data, so separating out the H. pylori was a tall order" (Milliken, 2016). Milliken explains that "the bacteria's genome contained a surprise. Instead of being like the modern European strain of H. pylori, Iceman's stomach bug was more closely related to one currently found in Northern India" (2016). The discovery of the now-extinct strain of H. pylori indicates that the "massive waves of migration into Europe occurred sometime after Ötzi died 5,300 years ago" (Milliken, 2016). Milliken describes H. pylori as "[a] microbe that can unlock a trove of secrets regarding the formation of Homo sapiens" (2016). Milliken optimistically ends their article by stating, "By scraping H. pylori from the wizened guts of ancient humanity, we can trace the path of modern human evolution and maybe even unlock secrets of human health relevant to people today" (2016).

Kean et al. "undertook a clinical musculoskeletal examination of the Iceman,...[using] photographs and radiographic imaging" (2013, p. 11). They first examined the "numerous linear carbon tattoos [on Ötzi's skin], which [were] not of a decorative type. These have been presumed to possibly be 'medicinal' tattoos administered for therapeutic reasons and may have been used in acupuncture-like treatment of pain" or they "may have been used for diagnosis or location of his painful states" (Kean et al., 2013, p. 11). Kean et al. then conducted "[s]pinal imaging" which "identified areas of spinal damage and [their] observations have provided clues as to possible sites of spinal initiated pain and hence sites for administration of the 'medicinal' tattoos" (2013, p. 11). Much like previous tests conducted by other researchers, Kean et al. discovered that Ötzi had arthritis (2013, p. 11). They also discovered "other forms of long-term musculoskeletal damage, but which do not have adjacent or corresponding 'medicinal' tattoos (Kean et al., 2013, p. 11). "[They also] contend that the back and leg 'medicinal' tattoos correspond directly to sites of chronic right knee and right ankle pain, and left thoracolumbar pain" (Kean et al., 2013, p. 11). "[The tattoos] also correspond to lower lumbar and sciatic referred radicular pain which may have a contributory cause related to the presence of a transitional lumbar 5 vertebra" (Kean et al., 2013, p. 11). Essentially, Ötzi had numerous tattoos on his back that corresponded with specific vertebrae, one on his right knee, numerous on his right ankle, and numerous on both of his shins (Kean et al., 2013, p. 13). Furthermore, "[t]he 'tattoo' site on the left wrist may not be a carbon tattoo but could have been a stain from leather, corresponding to the leather of the archer's wrist guard" (Kean et al., 2013, p. 14). Thus, "[i]f the 'medicinal tattoo' hypothesis is correct, this implies that the tattoo sites on the lower thorax, lumbar areas, posteromedial right knee, lower legs, and right ankle may be sites of chronic/persistent or recurrent pain" (Kean et al., 2013, p. 15). "[Ötzi's] belongings suggest he was an archer. If he adopted the right knee kneeling pose used by hunting archers this may have been the precipitant of chronic right knee pain related to ligament and/or meniscal damage" (Kean et al., 2013, p. 15). Additionally, "[w]hen the right knee kneeling position is assumed to shoot a long bow, the right ankle would be in the fully flexed and inverted position" (Kean et al., 2013, p. 15). So, "[i]f this position of the right knee and ankle was assumed frequently on a long-term basis (for hunting through his lifetime) it could account for chronic strain to the ankle ligaments" (Kean et al., 2013, p. 15). Kean et al. "conclude that the presence of the back and leg medicinal tattoos indicate that...[Ötzi] would [experience] episodes of chronic persistent mechanical back pain" along with "pain to the lower legs and possibly the right ankle due to referred pain and/or sciatic radicular pain of varying severity" (2013, p. 17).

Moreover, "[u]sing recent published data...of the genome structure of [Ötzi], we suggest...the osteoarthritis or inflammatory joint injury may relate to [the] presence of coronary heart disease (CHD) and Lyme disease (Borrelia burgdorferi) infection" (Kean et al., 2013, p. 11). In other words, "[t]he origins of his musculoskeletal conditions are unclear but there are indications that Lyme disease and CHD may have been factors" (Kean et al., 2013, p. 11). Kean et al. explain that "[a]therosclerosis and other vascular conditions are common in patients with osteoarthritis...thus perhaps the predisposition to coronary heart disease was contributory to some of the osteoarthritic joint damage" (2013, p. 18). "The genome sequencing of his DNA...also identified that he had been infected with the spirochete, Borrelia burgdorferi, the causative pathogen of Lyme disease" transmitted by ticks (Kean et al., 2013, p. 18). "Whether Ötzi was symptomatic...with some or several manifestation of Lyme disease, cannot be known, but evidence that he was infected with Lyme disease is an additional possible contributory factor to possible joint pain and joint damage" (Kean et al., 2013, pp. 18-19). Therefore, "[t]he presence of atherosclerosis and its association with osteoarthritis: infection with Lyme disease; and X-ray evidence of osteoarthritic joint abnormalities; and spinal damage" demonstrate "that Ötzi had joint and spinal pain... especially in his latter years" (Kean et al., 2013, p. 19).

Countless organizations have conducted research on Ötzi to understand all aspects of his life and body. The research ranges from numerous CT scans to extensive nanotechnological analyses of tissue samples (Zink et al., 2014, p. 204). Researchers have studied every inch of Ötzi's body including his "desiccated stomach" and his musculoskeletal system (Milliken, 2016; Kean et al., 2013, p. 11). The "genome approach" that Zink et al. conducted resulted in new facts about Ötzi including that he had brown eyes, was lactose intolerant, that he had arthritis, that his vascular calcification was a clinical marker for atherosclerosis, and that he had a genetic predisposition for coronary heart disease (2014, p. 205). Furthermore, Milliken explains that microbiologist Frank Maixner along with a team of scientists discovered an extinct strain of Helicobacter pylori in Ötzi's stomach which is related to a strain in present-day Northern India as opposed to the European strain (2016). Their discovery led to the further understanding that the large migration to Europe occurred after Ötzi's death (Milliken, 2016). Helicobacter pylori allows scientists to track the migration of humans around the world (Milliken, 2016).

Moreover, the musculoskeletal examination of Ötzi revealed interesting carbon tattoos along his back and legs (Kean et al., 2013, p. 11). The tattoos are believed to be medicinal and indicate where Ötzi was experiencing chronic pain (Kean et al., 2013, p. 11). The tattoos also indicate specific pain that Ötzi would have experienced due to his life as an archer where he would have been kneeling on his right knee and flexing his right ankle for long periods of time (Kean et al., 2013, p. 15). There is also a stain on his right wrist, potentially from leather, which would correlate with an archer's wrist guard (Kean et al., 2013, p. 14). Kean et al. contend that Ötzi's arthritis relates to his predisposition for coronary heart disease and his infection with Lyme Disease (2013, p. 11). While Kean et al. do not know if Ötzi was experiencing symptoms of Lyme Disease, they contend that the disease itself could have increased his joint pain and damage (2013, pp. 18-19). Despite the fact that extensive research has already been conducted, new technologies are arising daily and could offer even further insight into the life and body of Ötzi.

Chapter 8:
How has the Environment Affected the Ötzi Man? What Questions are We Still Asking About the Ötzi Man?

by Dasarathy Mutharasan

The environment around Ötzi Man had a monumental role in the preservation of his body and the many artifacts that were with him. Having lived around 5300 years ago (Dickson et al. 2003), his body was well protected and led to a greater understanding of the environment that led to his demise. This chapter will explore how the environment affected the Ice Mummy and is divided into two parts. The first part will briefly touch on the perfect conditions that allowed for his corpse to be uncovered thousands of years later in immaculate condition. The second part will explore how the environment shaped the Ötzi Man when he was very much alive. While we have learnt a lot from the Iceman, there are still many questions that remain. These will also be briefly touched on, throughout the chapter.

The Ice's Impact on Ötzi Man Post Mortem

The conditions that protected the Tyrolean Iceman were almost perfect. Researchers Bohleber et al. (2020) explored the ice at the unique site of the Austrian Alps as part of a research study. They specifically investigated the Weißseespitze summit glacier which was just 12 km from the site of the original Ötzi Man's discovery, which means it can be used as a potential proxy to understand the specific details of the death site of the Iceman. At this specific site, the researchers determined that the summit had pristine conditions that allowed for the ice to be preserved from thousands of years ago. This allows for some perspective to be gained on the perfect conditions that allowed us to extract the Ice Mummy and relevant historical information.

The researchers were able to analyze the ice gradient to determine the age of the glaciers. It revealed that the summit was likely ice-free or covered by smaller glaciers. It only became glaciated or covered in ice 5900 years ago, which is only approximately 500 years before the Iceman's time. The glaciation and formation of ice are what enabled his body to be preserved in an immaculate condition. Furthermore, Bohleber et al.'s (2020) research also revealed some unique preservatory conditions of the Austrian alps, especially at the summits. The limited thickness and dome geometry of the Weißseespitze summit meant that there was little ice flow at the ice divide. As Walther (2021) stated, while most glaciers flow down mountains, the glaciers at the summits form domes that do not move as much. This means that the ice at the Weißseespitze summit barely moved—preserving it for thousands of years. Overall, these conditions allow for an understanding of the many circumstances that led to the discovery of the Ötzi Man. Without the pristine environment's glaciation and the minimal glacial movement, we may have never uncovered Iceman who effectively became a time capsule for life 5000 years ago. The amount of historical information that was extracted from this mummy also has a great dependence on the environment that affected him while he was living.

The Environment's Impact on Ötzi Man Before His Death

The harsh living conditions of the environment could be gleaned off the items and the characteristics of the Iceman's discovery site. It can be argued that our understanding of the Tyrolean is a summary of the environment's impact on him, from thousands of years ago. His clothing, his tools, his food are all the result of the environment that he lived in. The environment's impact on him can be seen through the clothing and artifacts that were surprisingly well preserved. Researchers O'Sullivan et al. (2016) performed an extensive analysis of the many animal sources that were present in the clothing of the body. They found that the Tyrolean's coat was made of rudimentary sheep and goatskin stitched together. This indicates the dependence that individuals of the time had on domesticated animals, as he used whatever was available to him. His leggings were made of goat leather, coincidentally the same material as another discovery of 4500-year-old leggings from the Bernese Alps in Switzerland (O'Sullivan et al., 2016). All of these results indicate that humans at the time were heavily dependent on animals for resources inclusive of clothing. Furthermore, the selective choices of using distinct animal hides (goat leather for leggings, goat and sheep for his coat, etc.) indicate that individuals used these hides for certain characteristics like increased warmth and comfort.

At first, these might also indicate that he depended greatly on livestock farming for his fashion choices but his quiver and cap provide evidence of "hunting and trapping" wild animals (O'Sullivan et al. 2016). His cap contained genetic markers of brown bear and his quiver contained genetic markers of roe deer (O'Sullivan et al., 2016). Additionally, the materials of the quiver were held together by the strong, cord-like, plant-based bast material (Bortenschlager & Oeggl, 2000). This Tilia bast was the "best raw material available in his environment" as it was "the material of choice" in Central Europe. The bast could readily be used to produce cord rope and clothing.

While this may seem insignificant, this showed that the Iceman was picky in his material choices, as he had outstanding equipment at the time. To not have the best materials might have meant the difference between life and death in the harsh, cold environment. Other studies have indicated that the Ötzi man's clothing and artifacts contained traces of other species like wild canid (dog-like carnivores), red deer and Chamois (species of goat-antelope in Europe), something that researchers O'Sullivan et al. (2016) admitted ran contrary to their analysis of the Iceman's artifacts. Different sampling locations may explain this discrepancy, but it also raises some important questions as we may never know for sure that these species were used exactly in the manufacture of the Iceman's belongings. Nonetheless, the variety of domesticated and wild animal traces found in the Iceman's artifacts provide considerable evidence regarding the active lifestyle that he had to live. The hunting of wild species for clothing shows a lack of resources at the time as the Iceman was forced to forage and hunt for raw materials. In other words, he may have needed to adopt a hunting lifestyle to supplement his survival at the time. Proof of this was seen even in his last meals.

In another study, researchers Rollo et al. (2002) determined the last few meals that the Iceman had before his demise. Using DNA samples extracted from the site, it was determined that his last meal was red deer meat and cereal. His second to last meal was red ibex meat, dicot plants and cereals. A similar study by researchers Maixner et al. (2018) came to a similar conclusion regarding the last meals but also found that the meats were chemically different as they were "thermally treated" or cooked as protein denaturation was detected. As a previous study found that there were charcoal particles found in the man's intestinal tracts (Oeggl et al., 2007), this aligns with Maixner et al.'s (2018) conclusion regarding the heat treatment. Overall, the meals along with the clothing choices indicate the harshness of the environment as the Tyrolean Iceman had to rely heavily on opportunistic hunting and whatever natural resources near him to survive (ex: hunting a bear for his cap, consuming locally found grains, heating with charcoal). The environment had some negative impacts on the iceman. Due to the brutal and cold Alps environment, 3210m above sea level, the Iceman had to consume plenty of fatty meats to sustain his energy reserves and survive. Researchers Murphy et al. (2003) found that there were mineral deposits in his arteries and veins, evidence of cardiovascular health problems. Although Ötzi Man was genetically predisposed for cardiovascular disease (Keller et al., 2012), these high-fat meals are a risk factor for coronary heart disease, which may have contributed to atherosclerosis that he developed (Maixner et al. 2018).

The Iceman's health problems did not seem to end there. One of his fingernails exhibited 3 Beau's lines (Dickson et al., 2003) which occur when the nails stop growing abruptly and start growing again. They provide a vague medical history and showed that the Tyrolean had been severely ill thrice in just the last 6 months of his life. Furthermore, his last illness lasted 2 weeks—this was 2 months before his death (Dickson et al., 2003). Our inability to identify these illnesses brings up some important questions. If the Iceman led such an active lifestyle that was dependent on a mix of agricultural/hunter lifestyles, he surely needed some treatment to survive. The tough alpine environment must have contributed to the health problems he had, like the degenerative hip and knee joint problems. This highlights some important questions:

What was society's understanding of human health sciences? What diseases and illnesses were common at the time? We have a few clues that point to a rudimentary understanding of human health sciences at the time. Researchers Zink et al. (2019) delved into the possible evidence of medical treatment and care and found that the many tattoos of the Tyrolean Iceman could be therapeutic markers for acupuncture, which would have helped for the joint problems he had. Furthermore, many other herbs with medicinal properties were found on him. The high proportion of fern bracken and bog ferns found in his intestine along with the birch polypore found in his equipment are of particular interest. Fern bracken is a herbal medicine while bog fern is a medical dressing used for wounds. The birch polypore in his equipment is an active antibiotic substance. Researchers Zink and colleagues (2019) did acknowledge that this is not conclusive evidence as the ingestion of the herbs like fern bracken could have been unintentional ingestion, which highlights some historical knowledge gaps. Nonetheless, there was most likely some rudimentary understanding of medically beneficial plants and herbs at the time. Furthermore, it can be understood that the tattooist employed a considerable amount of effort into applying the 61 tattoos on his body (Zink et al., 2019) which lends credibility to the idea that there was some medical care present at the time. Nonetheless, many questions remain as we do not have much evidence regarding health science knowledge in the Copper Age. Understanding this along with the illnesses that afflicted Ötzi Man and possibly other individuals of his time would allow us to gain a better understanding of the human condition in the 4th millennium B.C.

Ötzi Man suffered even more abuse as it was discovered that his left 5th-9th ribs had broken and healed during his lifetime. Ötzi Man also had many other injuries but it was determined that this was due to "glacial action and rough recovery", meaning that he didn't sustain those injuries during his lifetime. While a ton of information has been gathered based on medical analyses of the Iceman, many questions remain. For instance, one researcher theorized that the Iceman was fleeing an attack in his home village. He even suffered a defensive wound where he used his hand to protect himself from his assailants but succumbed to his injuries in the Alps. However, many of the injuries that the Iceman had occurred after his death, which throws some serious doubt on this theory (Dickson et al., 2003). This highlights an important question: How did Ötzi Man die? We conclusively know that he had an arrowhead in his left shoulder blade that was fired at him from an attacker. This most likely led to his demise, but to determine that the attack came from his home village would be mere conjecture. Nonetheless, these clues of severe injury lend some considerable evidence of the tough conditions that humans had to endure at the time. It can also be understood that survival was prioritized before good health. Given the number of times he had to put himself at risk to get sick thrice within 6 months, the joint and muscle problems, the severe injury to his ribs and not to mention atherosclerosis (Dickson et al., 2003), the brutal environment had some deep ramifications on his health. Despite all of this, he also showed expertise and a strong understanding of the environment—this can be seen through the material choice for his clothing as he knew that warm animal skins and furs would protect him against the elements. This is expected because he needed to leverage the natural resources of the harsh environment he lived in—otherwise, he may have died much earlier. However, the natural environment was not the only thing that impacted the Tyrolean mummy as his demise could also be attributed to the impact of the social environment.

Popham (2013) reported on the emerging evidence from Dr. Tom Loy from Queensland University—where the Ice Mummy had traces of DNA from the blood of 4 different people—one on his cloak, one on his knife and 2 on the arrowheads that he carried. Furthermore, Ötzi died with "a flint-headed arrow" an inch away from his left lung. Dr. Loy theorized that the Ice Mummy was in a violent fight and was able to wound or kill his assailants severely. Ötzi Man began running home with a companion who tried to remove the arrow but broke it. The Mummy died as a consequence of his wounds. Again, this theory is far from conclusive as these are just a few clues that are part of a story that we will never uncover. While this narrative is again debatable, the clues of 4 distinct blood DNA traces along with the deadly arrow that most likely killed Ötzi Man provide a picture of the cruel social environment that the Tyrolean had to live in. Compared to today, there seemed to have been a lack of social order and law. As a consequence of this harsh social construct, the Italian mummy may have murdered others during his lifetime as evidenced by the blood traces —and was murdered himself as evidenced by the arrowhead in his corpse.

Closing Remarks

While many questions remain about Ötzi Man and society in the 4th millennium BC, the Tyrolean Ice Mummy has provided a substantial amount of historical information and clues regarding life 5000 years ago. The ice that conserved the Iceman must be acknowledged—the unique conditions of the Ötzal alps have allowed for the incredible preservation of his body and his tools which are of historic relevance. The icy conditions and the minimal ice flow ensured that his body would be conserved for thousands of years. The environmental impact of the ice cannot be understated as the Iceman went from just another unimportant human that lived in the Copper Age—to one of the most studied human bodies of all time.

Historians and researchers were able to gain a better understanding of the impact that the environment had on individuals like the Iceman. The alpine environment was brutal as the Iceman was forced to leverage the few resources at hand to survive. While he used domesticated animals like sheep and goatskin, he also utilized wild animals like deer and bear for his quiver and cap (O'Sullivan et al. 2016). The variety of materials used in the manufacture of his clothing indicate meticulous selection for certain furs and skins over another for advantages like warmth, comfort and so on. Furthermore, this also indicates that resources were far from plenty and that the Iceman needed to be resourceful by adopting a hunter and gatherer lifestyle to survive. This was seen even in his last few meals where he consumed ibex and deer meat along with dicot plants and cereal (Rollo et al., 2002). Considering the high energy required to traverse and survive in the Alps, the proportion of meat consumed is not surprising. Nevertheless, the active lifestyle and the tough environment took a substantial toll on him. The Iceman suffered from several health problems like atherosclerosis and muscle problems. While he was genetically prone to heart disease, the high fat and meaty diet that he survived on may have contributed to the atherosclerosis he had. He also had many muscle problems including joint and hip problems. With the many health issues he had, it seems likely that he needed some treatment to survive in the Alps—something that Zink et al. (2019) explored. The medical herbs found on him along with the potentially therapeutic tattoos on his body provide some evidence of a basic understanding of human health sciences. However, we still do not know whether he simply ingested the herbs accidentally. Lastly, the social environment was incredibly brutal as evidenced by the traces of blood from 4 distinct people found on him along with the arrowhead that most likely led to his demise (Popham, 2013). There have been some theories about a violent skirmish that the Iceman escaped, but we still do not know how exactly the Iceman sustained his injuries. Despite all the questions that remain, the Iceman has provided great insight into the human condition 5000 years ago—as it seems clear that the physical and social environment at the time was brutal for the Iceman.

Chapter 9:
What Controversy is There, Surrounding the Ötzi Man?
by Yash Joshi

Introduction

The Ötzi Man, also known as just Ötzi, has been accompanied with controversy ever since its discovery in 1991. Such a remarkable finding is bound to have multiple disputes, which have distinctively been seen over the years. One of the earliest issues surrounding the discovery of Ötzi was related to who had ownership over him. The body was discovered really close to the border of Austria and Italy, which led to disputes over which country had rights to Ötzi. Another major controversy surrounding Ötzi is connected to how the individual actually died. Ever since its discovery, there have constantly been new findings surrounding his death, but there has not been a single theory regarding the circumstances around his death which has been proven correct. The Ötzi Curse has also been a popular topic of conversation related to Ötzi as there has been a pattern that whichever researchers/people are associated with Ötzi, end up passing away soon after working with the mummy. While many people believe that the deaths are just a coincidence, there is a strong belief amongst people around the world that there is some sort of ancient curse attached to Ötzi.

Ownership of Ötzi

Ever since the naturally mummified body was found in the Ötzal Alps in 1991, there has been a fierce dispute regarding who owns the body. The conflict was between Italy and Austria, specifically during the time

Ötzi was at the Innsbruck University in Austria (Haller, 1998). Austria initially had custody of the mummy, but Italy had claimed that Ötzi had been found on Italian land and thus they had full rights to it. The Ötzi Man was specifically found in the Similaun Glacier in the Tirolean Ötztal Alps, which had shrunk in size and moved since the official boundary was declared between Austria and Italy in 1919 (Albeck-Ripka, 2016; The Editors of Encyclopaedia Britannica, 2020). Although the site at which Ötzi had been found drained into Austria, an official survey into the boundary between the countries revealed that the mummy was actually about 300 feet (or 100 m) into Italian territory (Albeck-Ripka, 2016).

After seven years of being studied at the university, Ötzi was transferred to South Tyrol, Italy where the Ötzi was going to be further studied and put on display, as decided by both countries (Rosenberg, 2020). Although this transfer was acknowledged by both countries, it was kept secret due to fears of retaliation from Austrian citizens. Specifically, there were many Austrian nationalists who never recognized the annexation of South Tyrol after World War 1, and they provided a risk for the transfer going through (Haller, 1998) At the South Tyrol Museum of Archaeology, Ötzi was placed in a special chamber to preserve its body and to be viewed by visitors. To remember the site where Ötzi resided for 5300 years, a stone marker was placed at the site of discovery (Rosenberg, 2020). Although there are currently no disputes regarding the ownership of Ötzi, for a period of time, tensions between both countries ran high.

The Death of Ötzi

Ever since the discovery of Ötzi, there have been numerous theories around how Ötzi actually came to its death. Evidence has continually been collected ever since the discovery and has led to various theories by researchers. Many theories have been refuted, while many others have been supported, but it is difficult to say for certain what theories regarding the death are correct. Studies have shown scientifically how Ötzi passed away, but the circumstances surrounding the death are unclear. The body was originally discovered by Helmut and Erika Simon while they were climbing. Helmut Simon's claim on discovering Ötzi's body was disputed by Slovenian actress Magdalena Mohar Jarc and Swiss hiker Sandra Nemeth, both who said they found the mummy first (Marks, 2005). The women became involved in a bitter battle with Simon and his wife over who found the mummy first, so much that Nemeth told the courts that she spat on Ötzi's body as DNA evidence (Marks, 2005). Jarc claimed that she saw Ötzi first and went to find a photographer, who was Simon (Marks, 2005). Nothing concrete became of these claims as Simon and his wife are still widely credited with the discovery of Ötzi.

The mummy had been intensively studied by researchers and after many years of study, x-rays showed that Ötzi passed away from an arrow-inflicted lesion to an artery near his left shoulder (Roach, 2007). Modern x-ray technology was used by Frank Rühli of the Institute of Anatomy at the University of Zurich in Switzerland, as well as colleagues from the General Hospital in Bolzano, Italy (Roach, 2007). From the x-ray examination, it was seen that the arrow had first penetrated the skin through the shoulder blade and got close to his lung, but did not penetrate the lung cavity, but rather stopped below the clavicle (Roach, 2007). Pathologist Eduard Egarter-Vigl and radiologist Paul Gostner of the Bolzano hospital have also shared that the arrowhead was 2 centimetres long and had actually shattered Ötzi's scapula (Holden, 2001). There was a ½ inch rip to the artery, where a hematoma (blood clot) was present in the visible tissue (Roach, 2007). The arrow had caused nerve damage and paralysis based on the location, meaning that Ötzi had a painful and long death where he bled out and suffered a cardiac arrest (Holden, 2001; Roach, 2007). The shaft was removed from Ötzi at the time of his death while he was still bleeding out, but the arrowhead remained inside Ötzi's back (Roach, 2007). Although these examinations helped researchers figure out how Ötzi lost his life, there is still major speculation about the circumstances surrounding the exact death.

Ötzi's Body

About his physical body, examinations found that his 12th ribs were missing and that his fifth to ninth ribs had broken and healed in the time that he was alive (Dickson et al., 2005). It was speculated that the multiple broken ribs could have been attributed to some incident during his life. There were also several indications that Ötzi was weak in health at the time of death. Most of his epidermis, hair, and nails were gone but there was still significant evidence present through the rest of his body. When examining one of his fingernails, researchers found three Beaus lines, which develop when nails stop and then start growing (Dickson et al., 2005). These lines were a clear indication that Ötzi had been ill three times in the last six months of his life, with the illness about two months prior to his death being the most severe (Dickson et al., 2005). Not only that, Horst Aspöck of the University of Vienna discovered that Ötzi had an infestation of an intestinal parasite whipworm, which can result in diarrhea or even dysentery (Dickson et al., 2005). Although these observations may not have resulted in his death, they explain that Ötzi probably was in a state of discomfort around the time of his death.

Even with the missing epidermis, there were still charcoal-dust tattoos visible on the layer of skin visible for Ötzi (Dickson et al., 2005). These marks are assumed to be therapeutic as many of them were on or close to Chinese acupuncture points and at places where Ötzi could have suffered from arthritis (Dickson et al., 2005). The x-ray examinations of Ötzi however show little if any sign of arthritis (Dickson et al., 2005). That has led to great speculation of what the tattoos are because if it was not arthritis, there must be some other explanation as to why Ötzi had these marks on his body. Other than that, Ötzi has very worn teeth which reflect his age (in the 40s) and diet, as well as frostbite on the little toe on his left foot (Dickson et al., 2005). Studies into Ötzi have noticed that he had many well documented sites of musculoskeletal damage such as osteoarthritis in the joints of the neck and right hip (Kean et al., 2013). The researchers reached the conclusion that Ötzi may have shown visible signs of right knee pain, right ankle pain, chronic mechanical back pain and sciatic nerve root irritation which include leg referred pain, leg and ankle weakness, leg tingling and numbness (Kean et al., 2013). Numerous x-rays, CT scans, examinations, and the locations of the tattoos helped the team reach these findings. They also believe that Ötzi might have walked with a limp at times of lower back and leg pain (Kean et al., 2013). It is unclear whether or not this pain was present at the time of his death, but there is a chance that his physical condition may have contributed to his death.

Theories about Ötzi's Death

One theory surrounding the death of Ötzi explains that he may have been killed as a sacrifice to the gods. The theory was brought forth by an archaeologist named Johan Reinhard who is an expert on various cultures of the Andes, the Himalaya, and other regions, as well as an authority on mummies and ritual sacrifices. He brought attention to the location of the arrowhead wound, which was on the back of Ötzi, speculating that it could have been murder or a ritual sacrifice (National

Geographic News, 2002). Ötzi's body was found in a naturally formed trench which is on a prominent pass between two of the highest peaks in the Otzal Alps. According to the archaeologist, this would be the type of location where populations from mountain cultures would make offerings to their mountain gods (National Geographic News, 2002). Mountain worship was a big part of cultures in prehistoric Europe in the Bronze Age, so there is a strong possibility that it was also a part of communities during the Copper Age, the time when Ötzi lived. As Ötzi was found buried in the trench, Reinhard believes that the body may have been buried in the trench by whoever killed him, hence why the body was so well preserved. A factor around the discovery of the body which baffled researchers trying to figure out the circumstances of the death related to why a copper axe was left with Ötzi's body. Reinhard believes that murderers would have taken the weapon if it was of use to them, but he rather thinks that the axe was deliberately left to help Ötzi in the afterlife or even as a tribute to the gods (National Geographic News, 2002). There is a possibility that the copper in the axe came from the mountains, which is why it would be a great token of appreciation to the gods. The theory surrounding the death of Ötzi presented by Reinherd does make sense, but has received skepticism from many researchers because of the lack of strong evidence that is necessary to prove this theory.

There are also contrasting ideas that stem from the arrowhead wound on Ötzi, where researchers believe that Ötzi died as a result of battle or a fight, known as the battle theory. The body of Ötzi was found alongside a plethora of weapons which included a dagger, bow and arrows, and an axe (Vince, 2002). These weapons were typical for an individual traveling through that area so that they would be able to protect themselves from predators or even potentially find some food. When the x-rays and microscopic imaging was done on Ötzi by Eduard Egarter-Vigl, the caretaker of Ötzi in the South Tyrol Museum of Archaeology in Bolzano, Italy, the examinations showed that the cut on the mummy's hand was deeper than originally expected (Vince, 2002). It was a 15mm-deep zig-zag cut to the hand, as well as a cut to his wrist, which Egarter-Vigl said was a result of defensive motions during a violent fight (Carroll, 2002; Vince, 2002). The cut in the right hand occurred when a sharp object like a flint-tipped spear or dagger punctured the base of his thumb, which in turn shredded his skin and muscle right to the bone (Carroll, 2002). The following blow hurt a bone in his wrist. The wound was believed to be fresh when he died because there was no scar (Carroll, 2002). Additionally, the discovery of the arrow blade in Ötzi's shoulder further suggested that his death was violent, almost ruling out the possibility of death because of drowning, hypothermia, or a fall which were brought forth by some early researchers (Carroll, 2002). The battle theory and the wounds on the arms of Ötzi suggest that Bronze Age tribes waged war on mountain peaks (Carroll, 2002). This would be significant as it would allow historians and researchers to better understand civilizations at that point in time.

Contrastingly, Konrad Spindler, an archaeologist from the University of Innsbruck, believed that the museum was incorrect and that Ötzi had not died a few hours after his injury (Vince, 2002). He believed that the injuries that Ötzi sustained were not life-threatening and he definitely died alone because all of his equipment was present and not stolen (Vince, 2002). Spindler went on to say that he thinks Ötzi was attacked on a ground he was familiar with and then fled his attacker into the Alps where he got surprised by early snowfall and froze to death. Spindler even came up with a whole theory about how exactly Ötzi got to the place of his death and died, known as the disaster theory. He proposed that Ötzi had fled into the mountains after getting injured in a fight in his home village and then he took refuge in the high pastures (Dickson et al., 2005). Being both hurt and tired, he fell asleep and later fell asleep on the boulder, where his body was preserved by snowfall (Dickson et al., 2005). This was one of the earlier theories proposed but has often been refuted because of later examinations which showed the severity of the injuries Ötzi sustained and how they contributed to his death. Also, while some early theories like those of Spindler have said that Ötzi might have been seeking shelter from bad weather in the trench but died there instead, researchers like Jason Reinhard have said that this does not seem reasonable. He shared that the trench is at a high point of the pass and not deep enough to provide any sufficient shelter, making it a poor place to seek refuge in a storm (National Geographic News, 2002). Once again, Spindler's theory does not seem reasonable, which explains why there has been such controversy regarding how Ötzi came to his death.

Eduard Egarter-Vigl wanted to examine the body for more evidence of a battle, in an attempt to solidify the battle theory. Alois Pirpamer, one of the earliest climbers who found Ötzi, shared how the knife was originally in Ötzi's right hand, but came loose when the body was pulled from the ice (Friend, 2003). This observation was ignored by Austrian authorities, which influenced many of the early theories about Ötzi's death. During Egarter-Vigl's analysis, he noticed there was a deep cut in Ötzi's right hand which had not been noticed in previous studies, along with a cut on his left hand and bruises on his torso, as if he had been beaten (Friend, 2003). After these observations, DNA specialist Thomas Loy of the University of Queensland in Brisbane, Australia was brought in to find microscopic blood samples on Ötzi which could help prove the death as a result of the battle theory. Blood from one individual was found on Ötzi's cloak, two more samples of blood were found on the arrow, and there was one more sample of blood on his knife, totalling to four additional samples of blood (Friend, 2003). These additional samples were all human DNA, strongly supporting the idea that some sort of battle occurred before Ötzi's death where Ötzi did damage to his attackers. Specifically, the investigative team believed that the blood on Ötzi's cloak might have come from a wounded friend that he may have been carrying, while the blood on the arrow might be the blood of people that Ötzi killed (Friend, 2003). Although the analysis showed evidence of other human blood, many researchers are still not fully convinced that this is the way that Ötzi passed away.

There has also been speculation that the loss of blood due to the arrow was not the cause of death for Ötzi. A team of researchers composed of prehistory professor Andreas Lippert of the University of Vienna, radiologists Dr. Paul Gostner and Dr. Patrizia Pernter from the Bolzano Regional Hospital, and Dr. Eduard Egarter-Vigl conducted some further research on Ötzi's body (Lorenzi, 2007). The team believed that the blood loss from the arrow would only have made Ötzi lose consciousness and that death actually came later from a blow to the head (Lorenzi, 2007). Death could have occurred via the attacker with a strike to the head with an object like a rock, or Ötzi could have fallen backwards and hit his head against a rock (Lorenzi, 2007). Regardless of how the death actually occured, the team concluded that cerebral trauma was the cause of death. Additionally, the team examined the unnatural position that Ötzi was found in. When found, Ötzi's body was face down with his left arm bent across his chest (Lorenzi, 2007). After the discovery of the arrowhead and the damage it caused, many believed that the position of the left arm was an attempt by Ötzi to stop the heamorrhage or acute pain caused by the arrow. However, with the realization that death was caused due to cerebral trauma, the researchers suggest that the unnatural position of

Ötzi's body could have been caused by the attacker after they attempted to pull out the arrow (Lorenzi, 2007).

In an analysis into Ötzi more recently, Detective Chief Inspector Alexander Horn has also provided insight on the death of Ötzi. He has specifically stated that it seems that Ötzi was caught by surprise by his attacker and that his attacker was about 30 m away when the arrow at Ötzi was shot (Bell, 2017). Not only that, Ötzi was relaxed up on the glacier as his bow was not ready for use and about half an hour before his death he was having a heavy meal, which are not signs of someone who was in a rush or fleeing (Bell, 2017). The injury sustained on his right hand seemed to happen a few days prior to his death, potentially during a fight because the wound had signs of a defensive injury (Bell, 2017). There is a possibility that this earlier fight was related to Ötzi death as that same person could have pursued him and killed him in the mountains, but that cannot be said with full certainty. Inspector Horn believes that the killing was personal and emotionally motivated as the attacker did not take the copper blade axe and other gear (Bell, 2017). Even he believes that there is a strong chance that the world will never be able to fully solve what happened to Ötzi.

The Ötzi Curse

A unique situation associated with the Ötzi is the curse that seems to befall those that are around the mummy. There is a belief that those that come into contact with the mummy end up dying of unique causes. Although it has not been seen with every person yet, there have been many incidences ever since Ötzi's discovery of researchers passing away. Helmut Simon, the German who found Ötzi while hiking in 1991, disappeared in the snow covered Alps with little hope of being found (Reuters in Vienna, 2004). Simon was hiking alone on the Garmskarhogel mountain in the Salzburg region, where there was a lot of snow (Reuters in Vienna, 2004). After the team composed of 93 men and search dogs in the area were unsuccessful, the search for Simon was suspended three days in and he was presumed dead (Reuters in Vienna, 2004). Simon was found dead eight days later as the victim of a 300 foot fall, with his body frozen and under a sheet of ice and snow (McMahon, 2005).

Additionally, Simon was not the only person who suffered the curse as Konrad Spindler, who was the head of the investigative team for Ötzi at Innsbruck University, died from complications related to multiple sclerosis (McMahon, 2005). He had spent years in the presence of the mummy studying it and had been aware of the curse theories, but disregarded them (McMahon, 2005). Dr. Rainer Henn, who placed Ötzi in the body bag died on his way to a conference to discuss his work with Ötzi (McMahon, 2005). The guide who organized the transport of the mummy by helicopter, Kurt Fritz, was killed in a snowslide in an accident in the mountains (McMahon, 2005). Surprisingly, the incident occured in an area that he knew well and he was the only climber out of the entire party that died. Rainer Hölz, a journalist who filmed the recovery of Ötzi, was a victim of brain tumour (McMahon, 2005). There is also a link to a possible sixth victim, Dieter Warnecke, the man who found Helmut Simon's dead body and then died of a heart attack after attending Simon's funeral (McMahon, 2005). Another victim of this Ötzi curse Dr. Thomas Loy, the scientist involved in finding the multiple blood samples on Ötzi which strongly pointed towards the battle theory of death of Ötzi. He was 63 years of age when he was found dead in his home in Brisbane, Australia and the autopsy for him came out inconclusive, meaning that he could have passed away of natural causes or an accident (Marks, 2005). Although the cause of death was not determined, Loy had not been in the best of health as 12 years prior to his death, around the time he got involved in the Ötzi project, he had been diagnosed with a hereditary disease which caused blood clots (Marks, 2005). All these deaths could just be attributed to natural causes, but it is still surprising that a lot of people who were involved with Ötzi are dead.

Conclusion

The discovery of Ötzi was monumental as it led to significant knowledge of populations that used to live in the past. Ever since the original discovery to this day, there seems to be a substantial amount of controversy and mystery associated with Ötzi. The ownership dispute between Austria and Italy was short and resolved in a civil manner as both countries were able to benefit from the discovery of Ötzi. The largest topic of discussion regarding Ötzi revolved around his death, as the circumstances surrounding the death as interpreted by researchers have constantly changed and further developed. From the initial disaster theory, to the many developments of the battle theory, further analysis on the mummy continues to reveal more information about how exactly Ötzi died, although it is near impossible to precisely determine the situations related to death. Although the idea is not acknowledged in scientific circles, the Ötzi Curse has become a topic that is prominent in relation to the Ötzi. There is a strong chance that all the deaths were just a coincidence, but it will be very difficult to convince some people that. Overall, Ötzi has brought forth a significant amount of uncertainty and continues to baffle researchers and the general public with the questions it leads to.

Chapter 10:
How has the Ötzi Man been Portrayed in Popular Culture?
by David Supina

The Ötzi iceman has been referenced a number of times by elements of popular culture, although he is featured more and less prominently depending on the specific work in question.

DJ Ötzi

It appears that Gerhard Friedle, better known by his stage name "DJ Ötzi", may have chosen his stage named after Ötzi, especially since Ötzi is just a shortening of the "Otztal Alps" in which the Ötzi iceman was found, and is not a meaningful word by itself. DJ Ötzi is probably best known in North America for his cover of the 1962 song "Hey Baby", although he has had a prolific career in German-speaking countries, where he has had numerous hits, including at least a dozen top ten albums in Austria (including two that hit number 1), four top ten albums in Germany, and five top ten albums in Switzerland. He also had ten songs in the top ten in Austria (and two that hit number 1), three number one songs in Germany and two others that hit the top ten, and three top ten songs in Switzerland (Idea Wiki, 2021).

5803 Ötzi Asteroid

There is an asteroid that has been named after Ötzi, that is between the orbits of Mars and Jupiter. Given the number of asteroids in our solar system, it's not surprising that one of them would be named after (Universe Guide, 2021).

Vom Neanderthaler Zum Ötzi

Vom Neaderthaler Zum Ötzi, which just means "From Neanderthal to Ötzi", is a fairly obscure card game that includes an Ötzi reference. Now, since this is a low profile game that seems like it has only ever been published in German, it is a little difficult to find out many details on it. On the board game database site Board Game Geek, the game has an entry, but it has an astonishing low single rating (that one person gave it a six out of ten) and only seven people on the website who claim to own it. It was apparently nominated for a 2006 award (which appears to be the year it came out) in a nine years and older category, and it does seem like it may have been intended to serve an educational purpose. One page on the internet, which seems to have originally hosted a link where you could buy and then download the game, seems to contain text juxtaposed against what appear to be cards from the game that describe the historical context of the game's contents (Doc Player, 2021). It is hard to tell how much the game would actually have to teach you about Ötzi, however, without a translated copy of the game or a German reader with the original, neither of which are, unfortunately, readily available. While there are likely thousands of card games that might have as engaging or better gameplay, it is interesting to see one with the name "Ötzi" in the title (Board Game Geek, 2021).

Ötzi Comic Books

There appear to be at least two Ötzi-themed comic books. One is simply entitled "Ötzi", which was written by Brennan Bajdek, and illustrated by Aleksander Bozic. It is a work of historical fiction, retelling the possible life of the Ötzi man, through his village being attacked and then sent on the run, reflecting on his life as he is chased down by pursuers. It is quite brief, running only twenty three pages, and is in English, although it may not contain much, if any, actual dialogue (ComiXology, 2021).

The other Ötzi comic appears to be in Spanish, called Ötzi. Por un puñado de ámbar, the subtitle which means approximately "For a Handful of Amber", credited as authored by Mikel Begoña and Iñaket (Iñaket looks like he has been credited as being the artist, Begoña as being responsible for the script) (Good Reads, 2021) and uses a much more abstracted art style. It proposes that Ötzi was an archer, and it appears to feature actual dialogue and generally includes a lot more text than the Ötzi project by

Bajdek and Bozic. It also appears to be spread out over three volumes, though it is difficult to tell if these three volumes together tell one complete story or if the project was abandoned (Norma Editorial, 2021). There appears to be another Ötzi comic, which does not seem to have much of a paper trail in English, credited to Bovo Eleonora and Barducci Armin (it's hard to tell if one or the other is the artist or the writer). From the description, it may focus less on the life of the Ötzi man himself, and the circumstances surrounding his relatively recent discovery (Athesia, 2019).

Who knew that the category of Ötzi themed comic books would have so many competitors?

Ötzi Novels

There appear to be a couple of novels themed around the Ötzi man. The first is called The Iceman: A Novel of Ötzi, written by Jeanne Blanchet, and released October 13th, 2020. It appears to be something of a speculative work of historical fiction, though it says in the description that the Ötzi man had a secret that could change his society permanently, which sounds like a flight of fancy to add drama to the life of the Ötzi man. The author acknowledges that she was hoping to create a historically accurate possibility, though she admits many of the exact details are unknown, in the author's notes. The prologue opens briefly with the modern day discovery of the Ötzi man (Amazon, 2021).

Another novel, written by Bud Seligson, was released November 8 2015, entitled Ötzi the Iceman. It seems like the focus of this novel is on the tribal religions of the time, and how Ötzi and his companions cross paths with a growing empire. It does appear to be more of an adventure thriller, judging from the description, though it's hard to tell if the author intended to follow history as studiously as the author of The Iceman. It is mentioned in the author's biography that he has a love of history, so it can be supposed that the contents of the novel's setting were not entirely fanciful (Amazon, 2021).

Ötzi the Cursed (Diablo III)

Ötzi also gets a reference in the popular video game, Diablo III. The full title of the creature is Ötzi the Cursed, Infected Creature of Death. It is termed a "unique accursed", which means it is both unique and an undead sort of creature that walks on all fours. It's design is humanoid in shape, but looks sickly and has long feet and fingers that point in spikes. But this brings up something that is worth mentioning, that isn't part of popular culture media, but is relevant to the Ötzi man's place in popular culture (Diablo Wiki, 2021).

The "Ötzi Curse"

There are some people who speak of an "Ötzi curse", linked to the idea of a Egyptian mummy's curse, since the Ötzi man is also mummified. There are a number of supposed strange "coincidences" surrounding those involved in the Ötzi man. Rainer Henn apparently died in a car crash on the way to give a speech about the Ötzi man. Kurt Fritz, who was a researcher on the Ötzi man, was killed in an avalanche only two years after the discovery of the Ötzi man. Helmut Simon, one of the two hikers who discovered the Ötzi man in the first place, had fallen to his death in the alps, despite being an experienced traverser of mountainous terrain. Dieter Warnecke, who led the search to find Helmut Simon after he disappeared on his hike, died of a heart attack mere hours after Helmut Simon's funeral.

Does this just seem like coincidence to you? Well, be careful about laughing this off, because that's what Konrad Spindler did, stating in an interview that he thought it was all just hype, and then he was the next person to die, in 2005, from complications due to multiple sclerosis. Maybe you should consider how ready you are to die before you try to laugh it off, then. But we're not done! Rainer Hölz filmed a documentary about Ötzi, but died of a brain tumour shortly after completing the film. Finally, to finish off the curse as it stands so far, Tom Loy did research on the Ötzi man's clothing, and then died due to a hereditary blood disease.

As to whether or not there is an actual curse, it's worth remembering that life is a condition that has, to date, a close to 100% mortality rate (there are all the people who are alive now that have yet to die, but other than that, the picture is pretty grim). If there were a substantial curse, why would there not be a more consistent pattern of how people have died? One would expect that if the Ötzi man were somehow haunting those who have dabbled with his remains, he would have maybe have given them the same violent death that he is suspected of having died from himself. Still, it is entertaining to consider (Ranker, 2021).

Animated Ötzi

There is an animated short film called Ötzi, produced by Evan Borja, for a Student Animation Festival by Cartoon Brew. The title seems to be a reference to the Ötzi man, although the plot of the short film is connected to the discovery only by virtue of being about discovering the remains of an ancient corpse in a snowy, mountain region. It takes a light tone as it involves the discoverer of the remains trying to extract the remains for a profit, only to accidentally destroy the remains. He then invents a time machine to go back in time to retrieve the corpse from before he destroyed it, only to accidentally overshoot his journey back in time to the last ice age and ends up becoming the corpse that he discovered (it's worth noting that by going back to the last ice age, this person's corpse would be over three times older than the actual Ötzi man). It is unclear whether by going back in time and becoming the corpse if he still destroys it in the future, but that may be too technical of a question for what is meant to be a light-hearted short. Before his untimely demise, the character does realize and regret not just selling the time machine for money. Hindsight is always twenty-twenty (Vimeo, 2012).

The Ötzi Man in Various Films

There are several films in which the Ötzi man plays a role, from documentaries to dramatizations.

The Iceman from Oetz Valley (Der Ötztalmann und seine Welt)

A 2000 film by writer and director Kurt Mundl and starring Arthur Burger, this is, much like the novels just discussed, a dramatization of a possible life of the Ötzi man, although it seems to be done in the style of a documentary (IMDB, 2021).

Horizon - Ice Mummies: Frozen in Heaven

Part of what appears to be a series of documentaries, Ötzi is one of several mummies covered in the episode, which was written by Tim Haines, and narrated by Nigel Le Vaillant. Interestingly, it features an appearance by Hilary Clinton (yes, the politician). This is one of many episodes on the series Horizon, season 33's eleventh episode (and apparently, 1124th overall! 1124th!), and it originally aired February 13th, 1997 (IMDB, 2021).

Ötzi e il mistero del tempo (Ötzi and the Mystery of Time)

If the first thing you thought when you thought of a frozen five thousand year old mummy was "great subject matter for a family or children's film", then you are in luck! This 2018 Italian film that describes a boy who has a passion for anthropology (do any of you remember being passionate about anthropology as a child?), who has his life set upside down when the mummified Ötzi springs to life, and then must hide Ötzi (which I guess is just his name?) from a pursuer. I am sure that hijinx will follow. It was directed by Gabriele Pignotta, and was credited to a writing team of Manuela Cacciamani, Carlo Longo, Giacomo Martelli and Davide Orsini. It runs for ninety minutes, and has, delightfully, a worldwide box office gross of $53,815. Perhaps Italian families just weren't ready for Ötzi to join the modern world (IMDB, 2021).

Iceman (Der Mann Aus dem Eis)

However, the most prominent feature on the Ötzi man is probably the feature film, simply called Iceman in its English release, the originally

German production called Der Mann Aus Dem Eis. It was released in 2017, and instead of featuring dialogue from a contemporary language like German or English, it had minimal dialogue in a language that was contemporary to the time and location of the Ötzi man, but featured no subtitles. It was written and directed by Felix Randau, and stars Jurgen Vogel, Andre Hennicke and Susanne Wuest. It describes the Ötzi man (named Kelab in the film) as the leader of a village, and has his settlement attacked (IMDB, 2021).

The film is described as fairly graphic and brutal, a revenge-filled tale that is relentlessly grim. It is also clearly meant to be period accurate (as you might have guessed that it uses a language of the Ötzi man's time, Rhaetic). The description mentions that Kelab may doubt his own actions as the film progresses towards his doom (I mean, you are sort of locked into one specific ending if you're making a film about the Ötzi man. It's somewhat inevitable) (Observer, 2019).

Conclusion

The Ötzi man has appeared in a large number of pieces of popular culture, ranging from mere references in the case of a musical performer, a comet, and a popular video game. However, there are several examples of media that contain much more extensive use of the Ötzi man as a focal point for consideration, whether they intend to be more objective in nature, like a documentary, or take more liberties like a comic book or a fictional movie, though even some of those clearly intend to portray the history of the time of the Ötzi man accurately. Much of the popular culture does appear to be not in English, at least initially, and it seems like the Ötzi man has more of a presence in Germany, Austria and Italy than in English speaking countries such as Canada and the United States. It may have even been possible to dig up further examples of popular culture references had I been operating in one of the aforementioned European countries. However, it is still interesting to see that the Ötzi man has had, over the thirty years since his initial discovery, clearly had an impact on the public consciousness. It might even be broader if we had looked to the "caveman frozen in ice" trope that has been used in popular culture at times, which owes some, though probably not all, of its lineage to the Ötzi man. But with the possible exception of the family film in which the Ötzi man has come back to life in modern day, there does seem to be some room for creativity in the use of the Ötzi man in more than just an exploration of society at the time of his life and death. He's a frozen man in ice; maybe we ought to stick him in a time machine.

Chapter 11:
What Future Direction will the Research and Analysis of the Ötzi Man Take?

by Ehimen Michael Ogadu

Introduction

The mystery behind the Ötzi man continues to puzzle scientists, researchers, scholars, and everyone who has studied the origin of this iceman who lived between 3400 and 3100 BCE. The iceman was discovered in September 1991 in the Ötztal Alps (hence the nickname Ötzi) which is the boundary between Austria and Italy. There exists pertinent questions surrounding his body, health, skeletal details and tattooing, clothes and shoes, tools and equipment, genetic analysis, blood and cause of death. This chapter seeks to discuss the forward thinking posture of the life and time of the Ötzi man.

The Descendants of the Ötzi man

Researchers for years have continued to delve into the lineage of the man believed to have been murdered in cold blood and for the past three decades questions around his lineage remain a mystery to all.

In addition, it is widely believed by scientists that the iceman had relatives. Unfortunately recent studies most certainly proves otherwise. Based on early analysis of the iceman's genes in 1994, many believed through the result, he was related to many living Europeans. This ideology was predicated on a limited sample of his genes. With improvement in technology scientists equipped with latest tools have recreated entire genome of the mitochondria — the tiny powerhouses within each cell — taken from intestinal cells of the remarkably well-preserved iceman(Richard Knox, Oetzi The Iceman May Be The Last In His Family, 2008).

A study carried out in 2008 in the November 1st issue of Current Biology reveals no similarity in mitochondrial genes from modern day humans collected. According to the National Human Genome Research Institute, Mitochondrial DNA is the small circular chromosome found inside mitochondria. The mitochondria are organelles found in cells that are the sites of energy production. The mitochondria, and thus mitochondrial DNA, are passed from mother to offspring. It is an important tool for constructing family trees and tracing the movement of people across time and space. Mitochondrial DNA has only about 16,500 genetic units, called base-pairs, instead of the 3 billion in the entire human genome. And the mitochondrial is peppered with lots of mutations, unique genetic tags that make it easier for scientists to track genetic lineages. Martin Richards, an archaeogenetics professor at the university of Leeds in his research "suspects that Öetzi, who was about 45 years old when he died, was at the end of his genetic line, or near the end. But he plans to seek samples of mitochondrial DNA from a thousand or so current denizens of the eastern Alps to be sure"Scientists are keen to know much more about the genealogy and history of the iceman. His existence suggests he lived in the Copper Age around 5300 years ago.

Europe was home to settled agricultural tribes, but anthropologists know little about the movements and cultures of people from that time. Some dating techniques employed by researchers to accurately predict his existence include;

Radiocarbon Dating

Radiocarbon dating, also known as carbon dating or carbon-14 dating is a method of determining the age of an object containing organic material, this technique was used in determining the age of Ötzi the iceman.

X-ray

An X-ray is a diagnostic test that uses radiation waves, called x-rays to take detailed photos of your body; this was used with the discovery of

Ötzi to determine whether Ötzi had any bone damage or underlying injuries. With the X-ray, archaeologists and historians were able to show that Ötzi had in fact died from a puncture wound in his shoulder which was untreated.

CT Scan

CT scan uses a computer that takes data from several x-rays that occur within a human or an animal's body, a CT scanner emits a series of narrow beams through the body and in turn displays numerous x-rays of the body, this was also used in the discovery of Ötzi the iceman.

On Going Research Projects

The south Tyrol museum of archaeology on its website has a range of scientific research projects with regards to the future direction the research and analysis of the Ötzi Man will take. These include;

Synopsis and retrospective evaluation of all radiological investigations of the Neolithic Iceman from 1991 to 2019

"All radiological investigations, including X-rays and computed tomography (CT) scans, that have been performed on the Iceman in the years since his discovery in 1991 are being digitally prepared and transferred to a collective archive. A subsequent comparison of the images is allowing the South Tyrol Museum of Archaeology to scientifically document the current state of the mummy. Since no other mummy in the world has been radiologically examined at such regular intervals, the long-term documentation of Ötzi's radiological images is also providing important comparative parameters for the conservation status of mummies in general".

Iceman Conservation Project 2.0

"Study regarding the conservation status of the mummy and evaluation of potential concepts for future conservation conditions of the Iceman and his associated finds. The most recent scientific findings and technologies will be taken into consideration. Project partner: EURAC-Institute for Mummy Studies"

Study of the Iceman's twisted cord

"Problem: detailed investigation of the cord bundle from Ötzi's quiver. Identification of the fiber type (plant- or animal-based), and test to determine if this cord could have been suitable as a bowstring".

"Unfreezing history. A study to find historical, technological and conservation possibilities for the earliest example of a Neolithic bow case ever to be found." (Swiss National Science Foundation (SNSF).

Flint tools and the contents of the belt pouch

"The contents of the Iceman's belt pouch and his flint tools underwent a new microscopic examination with the aim of determining the source of the raw material more precisely and analysing the signs of manufacture and use. The inventory concerned is special in that it is a closed complex, which, however, was not necessarily used or fashioned by one and the same individual.The project has been carried out in cooperation with the archaeologists Ursula Wierer, Simona Arrighi, Stefano Bertola and Jacques Pelegrin.

Result of the investigation: the origin of Ötzi's chert (flint) tools could be determined more precisely. According to this study, the raw material for his flint tools comes from the region between today's Veneto and the border with Lombardy. Ötzi does not seem to have been able to obtain new supplies for some time, as all his tools had been resharpened almost beyond their useful life. No doubt this circumstance was causing him a great deal of stress in the last few months and days of his life".

Manufacturing marks and signs of use on the axe blade

"The Iceman's axe is to undergo a new microscopic examination. The aim is to shed new light on the tool marks, for example the manufacture of the ridges and the pointed edge of the blade as well as the closure of the casting hole. The researchers will also focus on the signs of use, for instance, the impression of the angled haft on the neck of the axe blade and nicks on the blade.The project is carried out in cooperation with the Museum of Nature South Tyrol, Bolzano/Bozen, Italy".

Isotope analysis of the axe blade

"Investigations have been performed on the origin of the copper that was used to fashion the Iceman's axe. Analysis of the chemical composition and lead isotopes provided information about the site from which the raw material originated. Surprisingly, there was very little correlation between the copper in Ötzi's axe and copper found in Alpine deposits, which had been extensively studied in preliminary groundwork. Rather, the origin of the copper ore from the Iceman's axe clearly points to Southern Tuscany. The results were published in July 2017, with a supplement in December 2017.

The project is carried out with the Department of Geo-sciences, University of Padova, Italy".

Schnals – The high mountains as an economic and interactive area for prehistoric village communities

"This interdisciplinary research project aims to reconstruct how pastoral economies and highland agriculture functioned during the Bronze and Iron Age. It also aims to determine if the high mountains were territorially divided by settlers from Vinschgau (only in Schnals Valley)In recent years, archaeological investigations have been carried out at the entrance to Finailtal Valley and in Penaudtal Valley. In addition to sites from the middle Bronze Age (c. 1500 BC), structures from the Bronze and Iron Age were also found to have been used seasonally in the side valleys of Schnalstal, Finailtal, and Tisental. Schnals Valley's seasonal usage during the Bronze Age originated from Vinschgau and now it is important to determine which Bronze Age settlements were involved. Petrographic investigations of ceramic fragments from the Schnals high mountains and from settlements in Vinschgau should help to shed light on the valley's development.

In the past year surveys of the entire Schnals Valley revealed 13 more promising sites (4000 – 300 BC), some of which are being investigated in archaeological excavations. Up until now the only sites that had been known in this area were those that had been frequented by Mesolithic hunters. Interdisciplinary cooperation with palaeobotany (testing of pollen and macro-remains), archaeozoological investigations, and petrographic analyses of ceramic remains allowed conclusions to be drawn about the settlement and subsistence strategies.The research project is intended to last three years (2018-2020). It is a collaborative effort between the South Tyrol Museum of Archaeology, the Office of Archaeological Heritage of the Autonomous Province of Bolzano/South Tyrol, the Institute for Botany at the University of Innsbruck, and the Office of Geology and Building Materials of the Province Bolzano/South Tyrol"

Forensic analysis in cooperation with Munich Police Headquarters

"Numerous examinations have been performed since the Iceman was first discovered in 1991. A milestone in the history of research into the Iceman was the discovery of an arrowhead embedded in his left shoulder, which has shed new light on his story: the fact that Ötzi was murdered. The aims of the forensic analysis are to draw up a profile of the perpetrator, analyse the circumstances surrounding the Iceman's death and possibly determine the murder motive.First results were released in September 2016".

Determining the leather and hide samples found with Ötzi

"We know to a large extent which animals provided the hide and leather types discovered with the Iceman, but new research methods have permitted numerous corrections. Up until now, all skins and leather samples have been macroscopically determined by animal experts, while identification tests on a protein basis have also been carried out. For some samples it was possible to determine the animal family, but not the exact species. To confirm the first results from the University of Saarland (Germany), therefore, the DNA of all leather and fur samples from Ötzi has been examined in collaboration with the laboratory of the EURAC Institute for Mummy Studies in Bolzano. The results have been published in August 2016"

Scientists from the south Tyrol Museum of Archaeology In 2019 reconstructed the glacier conditions during the iceman lifetime. (Kelcie, 2019). The new research sheds more light on his immediate surroundings and what the mountain where he lived looked like. Researchers seem to agree this was around 5500 years ago. Those involved in the research, issued a publication December 2019 documenting their finding. Astonishingly, analysis of the Weißseespitze glacier, was carried out using state of the art technique. The Weißseespitze glacier was just a few kilometers from where the Ötzi man was first discovered. Scientists reconstructed an imagery of what was thought to be the original summits of the Ötztal Alps about a 1000 years ago which geographically is close to the border between Austria and Italy. Evidence from the research shows that the summit of the mountain remained glaciated throughout the Holocene(a time frame that represents about 11700 years ago).

Also, the site of the research at Weißseespitze was an ice dome in Austria and was never explored before. According to Pascal Bohleber, a fellow researcher on the team, he was quoted as saying "due to its dome-shaped geometry there is minimal to no ice flow at the ice divide."

Boundaries on a glacier between two regions of opposing flow directions are caused by ice divides. After careful analysis of the site at Weißseespitze, the results were amazing and fascinating. They began by drilling two large ice cores.

Bohleber further stated "We used state-of-the-art micro-radiocarbon ice dating . This technique uses the minimal amount of organic carbon that can be found in the ice, even in absence of any macroscopic organic fragments that you could see by eye. An even smaller fraction of this carbon is 14C, radiocarbon, which can be used to determine the age of the ice sample." From the dating and ice samples, scientists were able to calculate and make predictions as to what the ice covers looked like during the period of the iceman lived. Part of their observations were; from nearly ice free conditions high Alpine summits were emerging (bohleber, 2019)

In summary, "the results provide additional context for the change in ice cover that we see today," said Bohleber. "Although the current deglaciation of the summits during the Holocene may not be unprecedented, the pace may be. This is a topic on which we urgently need extensive empirical information."(bohleber, 2019)

Living Relatives Of Ötzi The Iceman?

There have been many questions surrounding the life of Ötzi The Iceman particularly on the subject of family. Did he have any and could we trace his ancestral lineage? Recent studies have attempted to find answers to these questions. To this end, there are claims of family members discovered via blood samples donated by 3,700 people in western Austria, not far from the Alps where Ötzi was found melting out of a glacier in 1991(draxler, 2013) Results from the DNA tests showed a unique genetic mutation from 19 donors with the mummified iceman. These men and the Iceman had the same ancestors.(Parson, 2013)

Furthermore, a sample was collected from the Ötzi man hipbone for genome sequencing a year prior. Scientist aimed at using these findings to trace his relationship to a regional ethnic group and most particularly discovering bloodlines from the iceman(parson, 2013)

It is believed his maternal line had become extinct hence researchers looked at DNA from the mummy's Y chromosome instead - genetic material passed down from father to child. This heritable, unchanging DNA was still found lurking in the genomes of Austrians alive today. Ötzi and his long-lost relatives fall into a rare European haplogroup and sub category (known as G-L91)(draxler, 2013)

Scientists and researchers alike still believe more studies need to be carried out as it is most likely certain there are more than 19 individuals with blood ties to the Ötzi man.

References

Chapter 1 References

Britannica, T. Editors of Encyclopaedia (2020, August 27). Ötzi. Encyclopedia Britannica. https://www.britannica.com/topic/Ötzi

Kutschera, W., & Rom, W. (2000). Ötzi, the prehistoric Iceman. Nuclear Instruments and Methods in Physics Research Section B: Beam Interactions with Materials and Atoms, 164-165, 12–22. https://doi.org/10.1016/s0168-583x(99)01196-9

Maderspacher, F. (2008). Ötzi. Current Biology, 18(21), 990–991. https://doi.org/10.1016/j.cub.2008.09.009

Püntener, A. G., & Moss, S. (2010). Ötzi, the Iceman and his Leather Clothes. CHIMIA International Journal for Chemistry, 64(5), 315–320. https://doi.org/10.2533/chimia.2010.315

Kutschera, W., Golser, R., Priller, A., Rom, W., Steier, P., Wild, E., Arnold, M., Tisnerat-Laborde, N., Possnert, G., Bortenschlager, S., &; Oeggl, K. (2000). Radiocarbon dating of equipment from the Iceman. The Iceman and His Natural Environment, 1–9. https://doi.org/10.1007/978-3-7091-6758-8_1

Wierer, U., Arrighi, S., Bertola, S., Kaufmann, G., Baumgarten, B., Pedrotti, A., Pernter, P., & Pelegrin, J. (2018). The Iceman's lithic toolkit: Raw material, technology, typology and use. PLOS ONE, 13(6). https://doi.org/10.1371/journal.pone.0198292

Drew, B. A. (2017). Tattoos in Medicine—From the Bronze Age to the

Modern Age. JAMA Dermatology, 153(2), 130. https://doi.org/10.1001/jamadermatol.2016.0224

Deter-Wolf, A., Robitaille, B., Krutak, L., &; Galliot, S. (2016). The world's oldest tattoos. Journal of Archaeological Science: Reports, 5, 19–24. https://doi.org/10.1016/j.jasrep.2015.11.007

Zink, A., Samadelli, M., Gostner, P., & Piombino-Mascali, D. (2019). Possible evidence for care and treatment in the Tyrolean Iceman. International journal of paleopathology, 25, 110–117. https://doi.org/10.1016/j.ijpp.2018.07.006

Handt, O., Richards, M., Trommsdorff, M., Kilger, C., Simanainen, J., Georgiev, O., Bauer, K., Stone, A., Hedges, R., Schaffner, W., Utermann, G., Sykes, B., & Pääbo, S. (1994). Molecular Genetic Analyses of the Tyrolean Ice Man. Science New Series, 264(5166), 1775–1778.

Muller, W. (2003). Origin and Migration of the Alpine Iceman. Science, 302(5646), 862–866. https://doi.org/10.1126/science.1089837

Kean, W. F., Tocchio, S., Kean, M., & Rainsford, K. D. (2013). The musculoskeletal abnormalities of the Similaun Iceman ("Ötzi"): clues to chronic pain and possible treatments. Inflammopharmacology, 21(1), 11–20. https://doi.org/10.1007/s10787-012-0153-5

Rollo, F., Ubaldi, M., Ermini, L., & Marota, I. (2002). Ötzi's last meals: DNA analysis of the intestinal content of the Neolithic glacier mummy from the Alps. Proceedings of the National Academy of Sciences, 99(20), 12594–12599. https://doi.org/10.1073/pnas.192184599

Maixner, F., Turaev, D., Cazenave-Gassiot, A., Janko, M., Krause-Kyora, B., Hoopmann, M. R., Kusebauch, U., Sartain, M., Guerriero, G., O'Sullivan, N., Teasdale, M., Cipollini, G., Paladin, A., Mattiangeli, V., Samadelli, M., Tecchiati, U., Putzer, A., Palazoglu, M., Meissen, J., … Zink, A. (2018). The Iceman's Last Meal Consisted of Fat, Wild Meat, and Cereals. Current Biology, 28(14), 2348–2355. https://doi.org/10.1016/j.cub.2018.05.067

Nerlich, A. G., Bachmeier, B., Zink, A., Thalhammer, S., & Egarter-Vigl,

E. (2003). Ötzi had a wound on his right hand. The Lancet, 362(9380), 334. https://doi.org/10.1016/s0140-6736(03)13992-x

Gostner, P., & Vigl, E. E. (2002). INSIGHT: Report of Radiological-Forensic Findings on the Iceman. Journal of Archaeological Science, 29(3), 323-326. https://doi.org/10.1006/jasc.2002.0824

Pernter, P., Gostner, P., Vigl, E. E., & Rühli, F. J. (2007). Radiologic proof for the Iceman's cause of death (ca. 5'300BP). Journal of Archaeological Science, 34(11), 1784-1786. https://doi.org/10.1016/j.jas.2006.12.019

Chapter 2 References

(www.dw.com), D. W. (n.d.). 'Curse of the Iceman' Linked to Scientist's Death: DW: 06.11.2005. DW.COM. https://www.dw.com/en/curse-of-the-iceman-linked-to-scientists-death/a-1765550#:~:text=Oetzi%20was%20discovered%20high%20in,being%20disturbed%20after%2053%20centuries.

Arie, S. (2004, October 24). 'Iceman' discoverer joins his find in Alpine grave. The Guardian. https://www.theguardian.com/science/2004/oct/24/germany.theobserver.

Barfield, L. (1994). The Iceman reviewed. Antiquity, 68(258), 10-26. https://doi.org/10.1017/s0003598x00046159

Bita, N. (2005, Nov 26). CURSE OF THE ICEMAN: [1 EDITION]. The Australian http://myaccess.library.utoronto.ca/login?qurl=https%3A%2F%2Fwww.proquest.com%2Fnewspapers%2Fcurse-iceman%2Fdocview%2F357440419%2Fse-2%3Faccountid%3D14771

Coakley, E. (2008, November 4). Oetzi the Iceman Has No Living Kin. findingDulcinea. http://www.findingdulcinea.com/news/science/2008/November/Oetzi-the-Iceman-Has-No-Living-Kin.htm.

Collins, R. (2016, October 3). This iceman's voice was reconstructed and heard again 5,000 years after his death. Irish Examiner. https://www.irishexaminer.com/opinion/columnists/arid-20423823.html.

Cullen, B. (2003, February 1). Testimony from the Iceman. Smithsonian. com. https://www.smithsonianmag.com/science-nature/testimony-from-the-iceman-75198998/.

Dickson, J. H. (2016). Ötzi, the Tyrolean Iceman. Encyclopedia of Geoarchaeology, 566–567. https://doi.org/10.1007/978-1-4020-4409-0_116

Ganesh, A. S. (2016, March 28). Meet Ötzi, the Iceman. The Hindu. https://www.thehindu.com/in-school/sh-science/meet-tzi-the-iceman/article7670928.ece.

Gibbons, A. (2018, August 28). Ötzi finders hit pay dirt, and scientists fret. Origins. https://blogs.sciencemag.org/origins/2009/06/otzi-finders-hit-pay-dirt-and-scientists-fret.html.

Gleirscher, P. (2014). Some remarks on the Iceman: his death and his social rank. Praehistorische Zeitschrift, 89(1). https://doi.org/10.1515/pz-2014-0004

Goodwin, S. (2004, Oct 25). Obituary: Helmut Simon; Finder of a Bronze Age man in the alpine snow: [First Edition]. The Independent http://myaccess.library.utoronto.ca/login?qurl=https%3A%2F%2Fwww.proquest.com%2Fnewspapers%2Fobituary-helmut-simon-finder-bronze-age-man%2Fdocview%2F310772555%2Fse-2%3Faccountid%3D14771

Heiss, A. G., & Oeggl, K. (2008). The plant macro-remains from the Iceman site (Tisenjoch, Italian–Austrian border, eastern Alps): new results on the glacier mummy's environment. Vegetation History and Archaeobotany, 18(1), 23–35. https://doi.org/10.1007/s00334-007-0140-8

Johnston, B. (2003, January 16). Couple who found iceman fight for compensation. The Telegraph. https://www.telegraph.co.uk/news/worldnews/europe/italy/1419057/Couple-who-found-iceman-fight-for-compensation.html.

Karnitschnig, M. (2004, Feb 03). The iceman cometh, but the trail of credit grows cold for his finders; as Austria, Italy tackle tyrolean tiff, the simons say fritz belongs to them. Wall Street Journal http://myaccess.library.utoronto.ca/login?qurl=https%3A%2F%2Fwww.proquest.com%2Fnewspapers%2Ficeman-cometh-trail-credit-grows-cold-his-finders%2Fdocview%2F308556978%2Fse-2%3Faccountid%3D14771

Levy, J. E. (2008, January 7). Iceman: Uncovering the Life and Times of a Prehistoric Man Found in an Alpine Glacier. AnthroSource. https://anthrosource.onlinelibrary.wiley.com/doi/full/10.1525/aa.2001.103.2.589.

McMahon, B. (2005, April 20). Scientist seen as latest 'victim' of Iceman. The Guardian. https://www.theguardian.com/science/2005/apr/20/science.italy.

Pilø, L. (2018, July 14). Ötzi – a new understanding of the holy grail of glacial archaeology. Secrets of the Ice. https://secretsoftheice.com/news/2018/07/04/otzi/.

Pringle, H. (2009, June 19). Beyond Stone and Bone " Money, Money, Money – Archaeology Magazine Archive. Beyond Stone and Bone RSS. https://archive.archaeology.org/blog/money-money-money/.

Rosenberg, J. (2020, August 27). Great Archaeological Discoveries of the 20th Century: Otzi the Iceman. ThoughtCo. https://www.thoughtco.com/otzi-the-iceman-1779439.

Simon, H., & Simon, E. (1991). photograph, Schnalstal/Val Senales Valley.

thelocal.de. (n.d.). https://www.thelocal.de/20090616/19991/.

Thumfart, W. F., Freysinger, W., Gunkel, A. R., & Truppe, M. J. (1997). 3D Image-guided Surgery on the Example of the 5,300-Year-Old Innsbruck Iceman. Acta Oto-Laryngologica, 117(2), 131–134. https://doi.org/10.3109/00016489709117753

Zink, A. R., & Maixner, F. (2019). The Current Situation of the Tyrolean Iceman. Gerontology, 65(6), 699–706. https://doi.org/10.1159/000501878

Chapter 3 References

Bradley, R. (2009). Image and audience rethinking prehistoric art . Oxford University Press.

Bradshaw Foundation. (2020, January 14). Facts about Altamira cave art. Bradshaw Foundation. https://bradshawfoundation.com/news/cave_art_paintings.php?id=Facts-about-Altamira-cave-art.

Chen, A. (2016, August 20). We finally know what Ötzi the iceman was wearing when he died 5,300 years ago. The Verge. https://www.theverge.com/2016/8/20/12560972/otzi-iceman-clothing-leather-copper-age.

Greshko, M. (2016, January 7). Iceman's Gut Holds Clues to Humans' Spread into Europe. National Geographic. https://www.nationalgeographic.com/culture/article/150107-otzi-iceman-stomach-microbes-science.

Kennedy, L. (2019, September 27). The Prehistoric Ages: How Humans Lived Before Written Records. History.com. https://www.history.com/news/prehistoric-ages-timeline.

Malone, C. (2003). The Italian Neolithic: A Synthesis of Research. Journal of World Prehistory, 17(3), 235-312. Retrieved May 23, 2021, from http://www.jstor.org/stable/25801207

Origjanska, M. (2018, January 27). Expert re-creates the shoes of 5,300-year-old Ötzi the Iceman, down to the bearskin soles and hay-stuffed lining. The Vintage News. https://www.thevintagenews.com/2017/12/31/otzi-the-iceman-shoes-2/.

Romey, K. (2016, August 18). Here's What the Iceman Was Wearing When He Died 5,300 Years Ago. National Geographic. https://www.nationalgeographic.com/culture/article/otzi-iceman-european-alps-mummy-clothing-dna-leather-fur-archaeology.

South Tyrol Museum of Archaeology. (n.d.). Clothing. Museo Archeologico dell'Alto Adige. https://www.iceman.it/en/clothing/.

Yeung, J. (2019, October 31). Frozen moss reveals fatal final journey of 5,300-year-old ice mummy. CNN. https://www.cnn.com/2019/10/31/europe/Ötzi-iceman-plants-study-intl-hnk-scli-scn/index.html.

Zink, A., Samadelli, M., Gostner, P., & Piombino-Mascali, D. (2019). Possible evidence for care and treatment in the Tyrolean Iceman. International Journal of Paleopathology, 25, 110–117. https://doi.org/10.1016/j.ijpp.2018.07.006

Chapter 4 References

Baroni, C., & Orombelli, G. (1996). The Alpine "Iceman" and Holocene Climatic Change. Quaternary Research, 46(1), 78–83. https://doi.org/10.1006/qres.1996.0046

Bohleber, P., Schwikowski, M., Stocker-Waldhuber, M., Fang, L., & Fischer, A. (2020). New glacier evidence for ice-free summits during the life of the Tyrolean Iceman. Scientific Reports, 10(1), 20513. https://doi.org/10.1038/s41598-020-77518-9

Bortenschlager, S., & Oeggl, K. (2012). The Iceman and his Natural Environment: Palaeobotanical Results. Springer Science & Business Media.

Emerson, D. (2019). Pyrite – the firestone. Preview, 2019(203), 52–64. https://doi.org/10.1080/14432471.2019.1696247

Ermini, L., Olivieri, C., Rizzi, E., Corti, G., Bonnal, R., Soares, P., Luciani, S., Marota, I., De Bellis, G., Richards, M. B., & Rollo, F. (2008). Complete Mitochondrial Genome Sequence of the Tyrolean Iceman. Current Biology, 18(21), 1687–1693. https://doi.org/10.1016/j.cub.2008.09.028

Gostner, P., & Vigl, E. E. (2002). INSIGHT: Report of Radiological-Forensic Findings on the Iceman. Journal of Archaeological Science, 29(3), 323–326. https://doi.org/10.1006/jasc.2002.0824

Gray, M. W. (1989). Origin and Evolution of Mitochondrial DNA. Annual Review of Cell Biology, 5(1), 25–50. https://doi.org/10.1146/annurev.cb.05.110189.000325

Green, R. E., Krause, J., Briggs, A. W., Maricic, T., Stenzel, U., Kircher, M., Patterson, N., Li, H., Zhai, W., Fritz, M. H.-Y., Hansen, N. F., Durand, E. Y., Malaspinas, A.-S., Jensen, J. D., Marques-Bonet, T., Alkan, C., Prüfer, K., Meyer, M., Burbano, H. A., ... Pääbo, S. (2010). A Draft Sequence of the Neandertal Genome. Science, 328(5979), 710–722. https://doi.org/10.1126/science.1188021

Hall, S. S. (2007). Last Hours of the Iceman. National Geographic, 4.

Hoogewerff, J., Papesch, W., Kralik, M., Berner, M., Vroon, P., Miesbauer, H., Gaber, O., Künzel, K.-H., & Kleinjans, J. (2001). The Last Domicile of the Iceman from Hauslabjoch: A Geochemical Approach Using Sr, C and O Isotopes and Trace Element Signatures. Journal of Archaeological Science, 28(9), 983–989. https://doi.org/10.1006/jasc.2001.0659

Ivy-Ochs, S., Kerschner, H., & Schlüchter, C. (2007). Cosmogenic nuclides and the dating of Lateglacial and Early Holocene glacier variations: The Alpine perspective. Quaternary International, 164–165, 53–63. https://doi.org/10.1016/j.quaint.2006.12.008

Kean, W. F., Tocchio, S., Kean, M., & Rainsford, K. D. (2013). The musculoskeletal abnormalities of the Similaun Iceman ("Ötzi"): Clues to chronic pain and possible treatments. Inflammopharmacology, 21(1), 11–20. https://doi.org/10.1007/s10787-012-0153-5

Keller, A., Graefen, A., Ball, M., Matzas, M., Boisguerin, V., Maixner, F., Leidinger, P., Backes, C., Khairat, R., Forster, M., Stade, B., Franke, A., Mayer, J., Spangler, J., McLaughlin, S., Shah, M., Lee, C., Harkins, T. T., Sartori, A., ... Zink, A. (2012). New insights into the Tyrolean Iceman's origin and phenotype as inferred by whole-genome sequencing. Nature Communications, 3(1), 698. https://doi.org/10.1038/ncomms1701

Keller, A., Kreis, S., Leidinger, P., Maixner, F., Ludwig, N., Backes, C., Galata, V., Guerriero, G., Fehlmann, T., Franke, A., Meder, B., Zink, A., & Meese, E. (2017). MiRNAs in Ancient Tissue Specimens of the Tyrolean Iceman. Molecular Biology and Evolution, 34(4), 793–801. https://doi.org/10.1093/molbev/msw291

Kutschera, W., Patzelt, G., Steier, P., & Wild, E. M. (2017). The Tyrolean Iceman and His Glacial Environment During the Holocene. Radiocarbon, 59(2), 395–405. https://doi.org/10.1017/RDC.2016.70

Magny, M., & Haas, J. N. (2004). A major widespread climatic change around 5300 cal. Yr BP at the time of the Alpine Iceman. Journal of Quaternary Science, 19(5), 423–430. https://doi.org/10.1002/jqs.850

Müller, W., Fricke, H., Halliday, A. N., McCulloch, M. T., & Wartho, J.-A. (2003). Origin and Migration of the Alpine Iceman. Science, 302(5646), 862–866. https://doi.org/10.1126/science.1089837

Murphy, W. A., Nedden, D. zur, Gostner, P., Knapp, R., Recheis, W., & Seidler, H. (2003). The Iceman: Discovery and Imaging. Radiology, 226(3), 614–629. https://doi.org/10.1148/radiol.2263020338

Oeggl, K. (2000). The diet of the Iceman. In S. Bortenschlager & K. Oeggl (Eds.), The Iceman and his Natural Environment: Palaeobotanical Results (pp. 89–115). Springer. https://doi.org/10.1007/978-3-7091-6758-8_8

Oeggl, Klaus, Kofler, W., Schmidl, A., Dickson, J. H., Egarter-Vigl, E., & Gaber, O. (2007). The reconstruction of the last itinerary of "Ötzi", the Neolithic Iceman, by pollen analyses from sequentially sampled gut extracts. Quaternary Science Reviews, 26(7), 853–861. https://doi.org/10.1016/j.quascirev.2006.12.007

O'Sullivan, N. J., Teasdale, M. D., Mattiangeli, V., Maixner, F., Pinhasi, R., Bradley, D. G., & Zink, A. (2016). A whole mitochondria analysis of the Tyrolean Iceman's leather provides insights into the animal sources of Copper Age clothing. Scientific Reports, 6(1), 31279. https://doi.org/10.1038/srep31279

Vincent, C., Meur, E. L., Six, D., & Funk, M. (2005). Solving the paradox of the end of the Little Ice Age in the Alps. Geophysical Research Letters, 32(9). https://doi.org/10.1029/2005GL022552

Zink, A., Samadelli, M., Gostner, P., & Piombino-Mascali, D. (2019). Possible evidence for care and treatment in the Tyrolean Iceman. International Journal of Paleopathology, 25, 110–117. https://doi.org/10.1016/j.ijpp.2018.07.006

Chapter 5 References

Dickson, J. H., Oeggl, K., & Handley, L. L. (2003). The Iceman Reconsidered. 288(5), 70–79. https://doi.org/10.2307/26060285

Hess, M. W., Klima, G., Pfaller, K., Ku"nzel, K. H., Ku"nzel, K., & Gaber, A. O. (1998). Histological Investigations on the Tyrolean Ice Man. In Am J Phys Anthropol (Vol. 106). https://doi.org/10.1002/(SICI)1096-8644(199808)106:4

Hollemeyer, K., Altmeyer, W., Heinzle, E., & Pitra, C. (2008). Species identification of Oetzi's clothing with matrix-assisted laser desorption/ionization time-of-flight mass spectrometry based on peptide pattern similarities of hair digests. Rapid Communications in Mass Spectrometry, 22(18), 2751–2767. https://doi.org/10.1002/rcm.3679

Kristensen, G. N. (2019). Original Paper A New Interpretation of Ötzi, the Iceman. 4(3). https://doi.org/10.22158/se.v4n3p165

Kutschera, W., & Rom, W. (2000). Ötzi, the prehistoric Iceman. Nuclear Instruments and Methods in Physics Research, Section B: Beam Interactions with Materials and Atoms, 164, 12–22. https://doi.org/10.1016/S0168-583X(99)01196-9

Maderspacher, F. (n.d.). Current Biology Magazine. Retrieved May 17, 2021, from http://www.archaeologiemuseum.it/index_ice.html

Püntener, A. G., & Moss, S. (2010). Ötzi, the iceman and his leather clothes. Chimia, 64(5), 315–320. https://doi.org/10.2533/chimia.2010.315

Rédei, G. P. (2008). Ice Man (Ötzi). In Encyclopedia of Genetics, Genomics, Proteomics and Informatics (pp. 952–952). Springer Netherlands. https://doi.org/10.1007/978-1-4020-6754-9_8170

Chapter 6 References

Dickson, J., Oeggl, K., and Handley L. (2003). The Iceman reconsidered. Scientific American, 288(5), 70–79. Retrieved May 18, 2021, from http://www.jstor.org/stable/26060285.

Hallett, V. (2018, Jan 22). Ötzi the Iceman died 5, 300 years ago, but he still needs regular checkups. The Washington Post. https://www.washingtonpost.com/lifestyle/kidspost/Ötzi-the-iceman-died-5300-years-ago-but-he-still-needs-regular-checkups/2018/01/22/4ccb66a0-f7d7-11e7-a9e3-ab18ce41436a_story.html.

Kutschera, W., and Rom W. (2000). Ötzi, the prehistoric Iceman. Nuclear Instruments and Methods in Physics Research Section B: Beam Interactions with Materials and Atoms, 164-165, 12-22. https://doi.org/10.1016/S0168-583X(99)01196-9.

Oeggl, K. (2009). The significance of the Tyrolean Iceman for the archaeobotany of Central Europe. Vegetation History and Archaeobotany, 18, 1-11. https://doi.org/10.1007/s00334-008-0186-2.

Püntener, A., and Moss S. (2010). Ötzi, the Iceman and his leather clothes. CHIMIA International Journal for Chemistry, 64(5), 315-320. https://doi.org/10.2533/chimia.2010.315.

Rosenberg, J. (2020, Jan 4). Ötzi the Iceman. ThoughtCo. https://www.thoughtco.com/Ötzi-the-iceman-1779439.

South Tyrol Museum of Archaeology. (2021, May 12). The Iceman. https://www.iceman.it/en/the-iceman/.

Walker, A. (2013, Mar 18). Appropriation (?) of the Month: Who should benefit from ancient human remains?: Legal, ethical, and economic challenges. Simon Fraser University. https://www.sfu.ca/ipinch/outputs/blog/appropriation-month-who-should-benefit-ancient-human-remains-legal-ethical-and-economic/.

Wikipedia. (2021, May 7). Ötzi. https://en.wikipedia.org/wiki/%C3%96tzi.
Yeung, J. (2019, Oct 31). Frozen moss reveals fatal final journey of 5,300-year-old ice mummy. CNN Travel. https://www.cnn.com/2019/10/31/europe/Ötzi-iceman-plants-study-intl-hnk-scli-scn/index.html.

Zink, A. R., and Maixner F. (2019). The current situation of the Tyrolean Iceman. Gerontology, 65(6), 699-706. https://doi.org/10.1159/000501878.

Chapter 7 References

Abedin, M., Tintut, Y., & Demer L.L. (2004). Vascular calcification. Arteriosclerosis, Thrombosis, and Vascular Biology, 24, 1161-1170. doi: 10.1161/01.ATV.0000133194.94939.42

Aghoghovwia, B. (2020). Subclavian artery and its branches. Kenhub. https://www.kenhub.com/en/library/anatomy/the-subclavian-artery-and-its-branches

Atherosclerosis. (n.d.). National Heart, Lung, and Blood Institute. https://www.nhlbi.nih.gov/health-topics/atherosclerosis

Cannon, J.W. (2018). Hemorrhagic shock. New England Journal of Medicine, 378, 370-379. doi: 10.1056/NEJMra1705649

Kean, W.F., Tocchio, S., Kean, M., & Rainsford, K.D. (2013). The musculoskeletal abnormalities of the Similaun Iceman ("Ötzi"): Clues to chronic pain and possible treatments. Inflammopharmacology, 21, 11-20. doi: 10.1007/s10787-012-0153-5

Maixner, F., Krause-Kyora, B., Turaev, D., Herbig, A., Hoopmann, M.R., Hallows, J.L., . . . Zink, A. (2016). The 5300-year-old Helicobacter pylori genome of the Iceman. Science (American Association for the Advancement of Science), 351, 162-165. doi: 10.1126/science.aad2545

Milliken, G. (2016). Iceman's stomach bug gives clues to humans' spread into Europe. Popular Science. https://www.popsci.com/icemans-gut-microbes-reveal-clues-early-human-geography/

Oxford University Press. (2021). Histology, n. Oxford English Dictionary. https://www-oed-com.ezproxy.library.ubc.ca/view/Entry/87291?redirectedFrom=histology#eid

Oxford University Press. (2021). Isotope, n. Oxford English Dictionary. https://www-oed-com.ezproxy.library.ubc.ca/view/

Entry/100187?redirectedFrom=isotope#eid

Oxford University Press. (2021). Paleobotany, n. Oxford English Dictionary. https://www-oed-com.ezproxy.library.ubc.ca/view/Entry/136168?redirectedFrom=paleobotany#eid

Spine-health. (n.d.). Degenerative arthritis definition. https://www.spine-health.com/glossary/degenerative-arthritis

Thomas, G.S., Wann, L.S., & Narula, J. (2014). What do mummies tell us about atherosclerosis? Global Heart, 9, 185-186. doi: 10.1016/j.gheart.2014.06.002

University of Pittsburgh Medical Center. (2018). What are vascular calcifications? UPMC HealthBeat. https://share.upmc.com/2018/09/what-is-vascular-calcification/

Zink, A., Wann, L.S., Thompson, R.C., Keller, A., Maxiner, F., Allam, A.H., . . . Krause, J. (2014). Genomic correlates of atherosclerosis in ancient humans. Global Heart, 9, 203-209. doi: 10.1016/j.gheart.2014.03.2453

Chapter 8 References

Bohleber, P., Schwikowski, M., Stocker-Waldhuber, M. et al. (2020). New glacier evidence for ice-free summits during the life of the Tyrolean Iceman. Sci Rep 10, 20513. Retrieved May 19, 2021 from https://doi.org/10.1038/s41598-020-77518-9

Bortenschlager, S., & Oeggl, K. (2000). The Iceman and his natural environment: palaeobotanical results. The Iceman and his Natural Environment. Springer. https://link.springer.com/book/10.1007/978-3-7091-6758-8

Dickson, J., Oeggl, K., & Handley, L. (2003). The Iceman Reconsidered. Scientific American, 288(5), 70-79. Retrieved May 19, 2021, from http://www.jstor.org/stable/26060285

Keller, A., Graefen, A., Ball, M., Matzas, M., Boisguerin, V., Maixner, F., Leidinger, P., Backes, C., Khairat, R., Forster, M., Stade, B., Franke, A., Mayer, J., Spangler, J., McLaughlin, S., Shah, M., Lee, C., Harkins, T. T.,

Sartori, A., Moreno-Estrada, A., ... Zink, A. (2012). New insights into the Tyrolean Iceman's origin and phenotype as inferred by whole-genome sequencing. Nature communications, 3, 698. https://doi.org/10.1038/ncomms1701

Maixner, F., Turaev, D., Cazenave-Gassiot, A., Janko, M., Krause-Kyora, B., Hoopmann, M. R., Kusebauch, U., Sartain, M., Guerriero, G., O'Sullivan, N., Teasdale, M., Cipollini, G., Paladin, A., Mattiangeli, V., Samadelli, M., Tecchiati, U., Putzer, A., Palazoglu, M., Meissen, J., Lösch, S., ... Zink, A. (2018). The Iceman's Last Meal Consisted of Fat, Wild Meat, and Cereals. Current biology : CB, 28(14), 2348–2355.e9. https://doi.org/10.1016/j.cub.2018.05.067

Murphy, W. A., Jr, Nedden Dz, D. z., Gostner, P., Knapp, R., Recheis, W., & Seidler, H. (2003). The iceman: discovery and imaging. Radiology, 226(3), 614–629. https://doi.org/10.1148/radiol.2263020338

Oeggl, K., Kofler, W., Schmidl, A., Dickson, J. H., Egarter-Vigl, E., & Gaber, O. (2007). The reconstruction of the last itinerary of "Ötzi", the Neolithic Iceman, by pollen analyses from sequentially sampled gut extracts. Quaternary Science Reviews, 26(7-8), 853–861. https://doi.org/10.1016/j.quascirev.2006.12.007

O'Sullivan, N. J., Teasdale, M. D., Mattiangeli, V., Maixner, F., Pinhasi, R., Bradley, D. G., & Zink, A. (2016). A whole mitochondria analysis of the Tyrolean Iceman's leather provides insights into the animal sources of Copper Age clothing. Scientific reports, 6, 31279. https://doi.org/10.1038/srep31279

Popham, P. (2013, October 26). DNA reveals how the Italian Iceman went down fighting. The Independent. https://www.independent.co.uk/news/world/europe/dna-reveals-how-italian-iceman-went-down-fighting-100039.html.

Rollo, F., Ubaldi, M., Ermini, L., & Marota, I. (2002). Ötzi's last meals: DNA analysis of the intestinal content of the Neolithic glacier mummy from the Alps. Proceedings of the National Academy of Sciences of the United States of America, 99(20), 12594–12599. https://doi.org/10.1073/pnas.192184599

Walther, K. (2021, January 13). Scientists Reconstruct the Glacial Conditions During Ötzi the Iceman's Lifetime. State of the Planet. https://news.climate.columbia.edu/2021/01/13/glacial-conditions-Ötzi-iceman/.

Zink, A., Samadelli, M., Gostner, P., & Piombino-Mascali, D. (2019). Possible evidence for care and treatment in the Tyrolean Iceman. International journal of paleopathology, 25, 110–117. https://doi.org/10.1016/j.ijpp.2018.07.006

References Chapter 9

Albeck-Ripka, L. (2016, June 26). Climate Change Is Shifting the Border Between Italy and Austria. https://www.vice.com/en/article/kwz539/melting-borders-v23n4

Bell, B. (2017, June 4). Who killed Oetzi the Iceman? Italy reopens coldest of cases. BBC News. https://www.bbc.com/news/science-environment-40104139

Carroll, R. (2002, March 21). How Oetzi the Iceman was stabbed in the back and lost his fight for life. The Guardian. http://www.theguardian.com/world/2002/mar/21/humanities.research1

Dickson, J. H., Oeggl, K., & Handley, L. L. (2005, January 1). The Iceman Reconsidered. Scientific American. https://doi.org/10.1038/scientificamerican0105-4sp

Friend, T. (2003, November 8). "Iceman" was murdered, science sleuths say. https://usatoday30.usatoday.com/news/health/2003-08-11-iceman-murder_x.htm

Haller, V. (1998, April 19). Bronze Age Iceman On Display In Italy. https://www.seattletimes.com/
Holden, C. (2001, July 26). Ice Man Was Killed From Behind. Science | AAAS. https://www.sciencemag.org/news/2001/07/ice-man-was-killed-behind

Kean, W. F., Tocchio, S., Kean, M., & Rainsford, K. D. (2013). The musculoskeletal abnormalities of the Similaun Iceman ("Ötzi"): Clues to chronic pain and possible treatments. Inflammopharmacology, 21(1), 11–20. https://doi.org/10.1007/s10787-012-0153-5

Lorenzi, R. (2007, August 31). Blow to head, not arrow, killed Ötzi the iceman—ABC perth—Australian Broadcasting Corporation. https://www.abc.net.au/science/articles/2007/08/31/2020609.htm?site=perth&topic=health

Marks, K. (2005, November 5). Curse of the Oetzi the Iceman strikes again—Independent Online Edition > Europe. https://web.archive.org/web/20070518151436/http://news.independent.co.uk/europe/article324955.ece

McMahon, B. (2005, April 20). Scientist seen as latest "victim" of Iceman. The Guardian. http://www.theguardian.com/science/2005/apr/20/science.italy

National Geographic News. (2002). Did "Iceman" of Alps Die as Human Sacrifice? 1.

Reuters in Vienna. (2004, October 19). Iceman's finder missing. The Guardian. http://www.theguardian.com/world/2004/oct/19/austria

Roach, J. (2007, June 7). Iceman Bled Out From Arrow Wound, X-Ray Scan Reveals. Science. https://www.nationalgeographic.com/science/article/-iceman-bled-out-from-arrow-wound--x-ray-scan-reveals

Rosenberg, J. (2020, January 4). Great Archaeological Discoveries of the 20th Century: Ötzi the Iceman. ThoughtCo. https://www.thoughtco.com/Ötzi-the-iceman-1779439

The Editors of Encyclopaedia Britannica. (2020, August 27). Ötzi | Discovery & Facts. Encyclopedia Britannica. https://www.britannica.com/topic/Ötzi

Vince, G. (2002, March 21). Iceman "died after knife fight." New Scientist. https://www.newscientist.com/article/dn2072-iceman-died-after-knife-fight/

References Chapter 10

Athesia. (2019). Ötzi The Iceman - Der Mann aus dem Eis - L'uomo venuto dal ghiacciaio. Athesia. https://www.athesiabuch.it/item/20227090

Amazon. (2021). The Iceman: A Novel of Otzi. Amazon. https://www.amazon.com/Iceman-Novel-Otzi-Jeanne-Blanchet/dp/1977231101

Amazon. (2021). Otzi the Iceman. Amazon. https://www.amazon.ca/dp/B07112W1WQ/ref=dp-kindle-redirect?_encoding=UTF8&btkr=1

Board Game Geek. (2021). Vom Neanderthaler zum Ötzi. Board Game Geek. https://boardgamegeek.com/boardgame/33039/vom-neanderthaler-zum-otzi

ComiXology. (2021). Ötzi. ComiXology. https://www.comixology.com/%C3%96tzi/digital-comic/456650

Diablo Wiki. (2021). Otzi the Cursed. Diablo Wiki. https://diablo.fandom.com/wiki/Otzi_the_Cursed

Doc Player. (2021). Sie können das Spiel Vom Neanderthaler zum Ötzi auf unseren Internetseiten unter. Doc Player. https://translate.google.com/translate?hl=en&sl=de&tl=en&u=https%3A%2F%2Fdocplayer.org%2F61102401-Sie-koennen-das-spiel-vom-neanderthaler-zum-oetzi-auf-unseren-internetseiten-unter.html&anno=2&prev=search

Good Reads. (2021). Ötzi. Por un puñado de ámbar. Good Reads. https://www.goodreads.com/book/show/32079146-tzi-por-un-pu-ado-de-mbar

Idea Wiki. (2021). DJ Ötzi. Idea Wiki. https://ideas.fandom.com/wiki/DJ_%C3%96tzi

IMDB. (2021). Der Mann Aus Dem Eis. IMDB. https://www.imdb.com/title/tt5907748/?ref_=ttpl_pl_tt
IMDB. (2021). "Horizon" Ice Mummies: Frozen in Heaven. IMDB. https://www.imdb.com/title/tt0134249/?ref_=kw_li_tt

IMDB. (2021). The Iceman from Oetz Valley. IMDB. https://www.imdb.com/title/tt0233008/

IMDB. (2021). Ötzi e il mistero del tempo. IMDB. https://www.imdb.com/title/tt6642926/?ref_=ttpl_pl_tt

Norma Editorial. (2021). ÖTZI. POR UN PUÑADO DE ÁMBAR. Norma Editorial. https://www.normaeditorial.com/ficha/comic-europeo/oetzi/oetzi-por-un-punado-de-ambar

Observer. (2019). The Grim Epic 'Iceman' Proves That Life Was Truly Miserable for the World's Oldest Mummy. Observer. https://observer.com/2019/03/iceman-review-rex-reed/

Ranker. (2021). Scientists Found A Perfectly Preserved Ice Corpse - And Then, One By One, They Died. Ranker. https://www.ranker.com/list/otzi-iceman-curse/april-a-taylor

Universe Guide. (2021). (5803) Otzi Asteroid. Universe Guide. https://www.universeguide.com/asteroid/9012/otzi

Vimeo. (2012). Otzi. Vimeo. https://vimeo.com/45228376

References Chapter 11

Püntener, A. G., & Moss, S. (2010). Ötzi, the Iceman and his Leather Clothes. CHIMIA International Journal for Chemistry, 64(5), 315–320. https://doi.org/10.2533/chimia.2010.315

Vicky Hallett (2018). Ötzi the Iceman died 5,300 years ago, but he still needs regular checkups https://www.washingtonpost.com/lifestyle/kidspost/otzi-the-iceman-died-5300-years-ago-but-he-still-needs-regular-checkups/2018/01/22/4ccb66a0-f7d7-11e7-a9e3-ab18ce41436a_story.html

Oliver Peschel (2019). The south Tyrol museum of archaeology https://www.iceman.it/en/the-research/ongoing-research-projects/

Saarland University (2019). "Ötzi the Iceman: Researchers validate the stability of genetic markers." https://www.sciencedaily.com/releases/2017/02/170216094511.htm

Observer. (2019). The Grim Epic 'Iceman' Proves That Life Was Truly Miserable for the World's Oldest Mummy. Observer. https://observer.com/2019/03/iceman-review-rex-reed/

Observer. (2019). The Grim Epic 'Iceman' Proves That Life Was Truly Miserable for the World's Oldest Mummy. Observer. https://observer.com/2019/03/iceman-review-rex-reed/

Levy, J. E. (2008, January 7). Iceman: Uncovering the Life and Times of a Prehistoric Man Found in an Alpine Glacier. AnthroSource. https://anthrosource.onlinelibrary.wiley.com/doi/full/10.1525/aa.2001.103.2.589.

McMahon, B. (2005, April 20). Scientist seen as latest 'victim' of Iceman. The Guardian. https://www.theguardian.com/science/2005/apr/20/science.italy.

Pilø, L. (2018, July 14). Ötzi – a new understanding of the holy grail of glacial archaeology. Secrets of the Ice. https://secretsoftheice.com/news/2018/07/04/Ötzi/.

Pringle, H. (2009, June 19). Beyond Stone and Bone " Money, Money, Money - Archaeology Magazine Archive. Beyond Stone and Bone RSS. https://archive.archaeology.org/blog/money-money-money/.

Rosenberg, J. (2020, August 27). Great Archaeological Discoveries of the 20th Century: Ötzi the Iceman. ThoughtCo. https://www.thoughtco.com/Ötzi-the-iceman-1779439.

Simon, H., & Simon, E. (1991). photograph, Schnalstal/Val Senales Valley thelocal.de. (n.d.). https://www.thelocal.de/20090616/19991/.

Walther .k(2021, Jan 13) Scientists Reconstruct the Glacial Conditions During Ötzi the Iceman's Lifetime https://news.climate.columbia.edu/2021/01/13/glacial-conditions-otzi-iceman/

Draxler(2013,Oct 16) Living Relatives Of Ötzi The Iceman Mummy Found In Austria
https://www.discovermagazine.com/planet-earth/living-relatives-of-otzi-the-iceman-mummy-found-in-austria

www.ingramcontent.com/pod-product-compliance
Lightning Source LLC
Chambersburg PA
CBHW030120170426
43198CB00009B/682